Fernanda de Oliveira S. Ferrari

Phosphogypsum for the production of construction materials

Fernanda de Oliveira S. Ferrari

Phosphogypsum for the production of construction materials

ScienciaScripts

Imprint
Any brand names and product names mentioned in this book are subject to trademark, brand or patent protection and are trademarks or registered trademarks of their respective holders. The use of brand names, product names, common names, trade names, product descriptions etc. even without a particular marking in this work is in no way to be construed to mean that such names may be regarded as unrestricted in respect of trademark and brand protection legislation and could thus be used by anyone.

Cover image: www.ingimage.com

This book is a translation from the original published under ISBN 978-3-330-76357-9.

Publisher:
Sciencia Scripts
is a trademark of
Dodo Books Indian Ocean Ltd. and OmniScriptum S.R.L publishing group

120 High Road, East Finchley, London, N2 9ED, United Kingdom
Str. Armeneasca 28/1, office 1, Chisinau MD-2012, Republic of Moldova, Europe
Managing Directors: Ieva Konstantinova, Victoria Ursu
info@omniscriptum.com

Printed at: see last page
ISBN: 978-620-8-40495-6

DEDICATORY

To my husband, Eduardo

To my parents, Ronaldo and Beatriz

To my sister, Patrícia To my brother-in-law, Ézio To my family and friends.

ACKNOWLEDGEMENTS

I would like to thank the coordinators, teachers and staff of the Interdisciplinary Postgraduate Programme in Engineering (PIPE) for their commitment, dedication and attention.

I would like to thank my supervisor, Vsévolod Mymrine, who always taught and shared a great deal of his knowledge and experience.

I would like to thank my parents, my sister and my husband who have always supported me in this and every stage of my life.

I would like to thank the team at the Minerals and Rocks Analysis Laboratory (LAMIR), Rodrigo, Evelin, José and Franciele, for their help with the analyses during the research.

I would like to thank Prof Dr Haroldo Araújo Pontes for the use of the Environmental Technology Laboratory (LTA) at the Federal University of Paraná (UFPR) and, above all, for his support and guidance during the final stage of the work.

I would like to thank Dr Nice Kaminara for her prompt assistance in developing the graphs.

I would like to thank the Scanning Electron Microscopy Laboratory at UFPR and BOSCH for their kindness and for the analyses carried out.

To everyone who helped me, indirectly or directly, during the entire period of this work.

SUMMARY

This study was dedicated to the use of phosphogypsum - waste from the production of phosphoric fertiliser, lime production waste and gold mining sand to form new construction materials. The results obtained in the uniaxial compressive strength tests reached 8.3 MPa on the 3rd day of curing and 13.5 MPa on the 90th day. The material had a water resistance coefficient of 0.95 at 28 days, water absorption of 8.6% and expansion values at day 3 of 1% and 1.5% after a year and a half of curing. The mechanical properties of the materials developed met the requirements of NBR 7.170/83 for solid bricks in classes A, B and C and ceramic blocks in classes 15 and 25. Using XRD, XRF, SEM and EDS, the physico-chemical processes involved in the formation of structures in the new materials were investigated, due to the transformation of crystalline minerals and the synthesis of amorphous substances, similar to the structures present in gypsum.

Key words: Phosphogypsum; Gold mining sand; Lime production waste; New construction materials.

SUMMARY

CHAPTER 1

INTRODUCTION

There are at least two characteristics common to both industrial production and various anthropogenic activities: the waste of raw materials and energy, which generally results in the intense generation of waste (LEMOS, 1998).

Phosphogypsum (PG) is the waste generated during the production process of phosphoric acid - P2O5, whose main raw material is apatite rock (AQUINO, 2005). FG has similar physical and chemical properties to gypsum (CANUT, 2005). In Brazil, to date, there are no specific rules regulating the use of GF, so a large volume of this waste is stored uncovered at the production site and only a small portion is reused (VILLAVERDE, 2008). The increase in the production of GF and the need for ever larger areas for its storage have fuelled interest in recycling this industrial waste. The main applications of GF are in agriculture, cement production and construction, which is the subject of this study.

In gold extraction, as in other mining activities, the topsoil is removed which, after the extraction process, generates, among other materials, gold extraction sand (OES) (CAHETÉ, 1995). As well as being used to rebuild the roof after extraction, OES can also be reused in the construction industry.

Research carried out in northern Vietnam to obtain new materials using industrial waste used lime production waste powder (RPC) with low-quality Portland cement to produce a cement composite, obtaining a high-performance material (STROEVEN *etal*, 2001).

In this context, this work is dedicated to trying to solve the problems currently encountered in the process of final disposal of the above-mentioned solid waste, especially phosphogypsum, due to its relevance from the point of view of the volume of generation, with the aim of creating a new final disposal alternative.

1.1 Work Objectives

1.1.1 General objective

Developing new materials for civil construction using phosphogypsum, lime production waste and gold mining sand, reducing the risk of impacts on the environment , by properly disposing of this waste.

1.1.2 Specific objectives

i. Characterise the materials in order to determine their potential for use as components of a new material;
ii. Developing technologies for the production of materials at laboratory and pilot plant level;
iii. Carry out tests to evaluate the properties of new materials.

CHAPTER 2

LITERATURE REVIEW

2.1 Plaster

Gypsum is a very complex product that has different phases depending on the thermodynamic conditions in which they are formed (ULMANSS, 1985).

Gypsum is one of the best-known materials used by humans. It is an inorganic binder, just like lime and cement, which means that although it needs water to harden, once it has taken its final form, gypsum does not resist the action of water. Although the terms "gypsum", "gypsum" and "gypsum" are used synonymously, gypsum refers to the mineral in its natural state, while gypsum refers to the calcined material (ABREU, 1965).

Gipso can be found in sedimentary soils from different geological periods, occurring most frequently in the Cretaceous and Tertiary. The origin of the large masses of gypsum contained in sedimentary rocks is attributed to the evaporation of ancient seas, but these minerals can also be formed, in special cases, by the action of gases and sulphuric waters acting on limestone (ABREU, 1965). Gypsum is basically composed of gypsum, anhydrite and some impurities, usually clay minerals, calcite, dolomite and organic material (CINCOTTO eia/, 1988).

Gypsum hydration is a chemical phenomenon in which the anhydrous material in powder form is transformed into a dihydrate due to a chemical reaction of the powder with water, according to the chemical reaction described in equation 2.1 (HINCAPIÉ; CINCOTTO, 1995).

$$CaSO_4.0,5H_2O + 1,5H_2O \rightarrow CaSO_4.2H_2O + CALOR \qquad (2.1)$$

2.1.1 Phosphogypsum

Phosphogypsum is the waste generated during the phosphoric acid production process, whose main raw material is apatite phosphate rock, considered an economically viable source of phosphorus for the production of phosphate fertilisers and phosphates for other purposes (AQUINO, 2005).

2.1.1.1 Phosphogypsum generation

For the manufacture of phosphate fertilisers, apatite, Ca3(PO4)2(OH, F, Cl), is the main raw material, which in the industrial process is attacked with sulphuric acid, with this process, the elements of the matrix rock are redistributed in the intermediate, final and by-product products. The products obtained by the fertiliser industry are phosphoric acid, simple superphosphate (SSP), triple superphosphate (TSP), monoammonium phosphate (MAP) and diammonium phosphate (DAP) (MAZZILLI, 2005).

Phosphogypsum is a by-product of the phosphate fertiliser industry, generated during the production of phosphoric acid by reacting phosphate rock with sulphuric acid. In Brazil, for economic reasons and adaptation of the industrial process, this is the process most commonly used in the production of phosphoric acid (VILLAVERDE, 2008). It is estimated that 4.5 to 5.5 tonnes of phosphogypsum are generated for every tonne of phosphoric acid produced. Relatively little of the by-product is used, and annual world production is approximately 150 million tonnes, which causes severe environmental impacts (CHANG *etal*, 1990).

2.1.1.2 Production of phosphoric acid for phosphate fertilisers

Phosphate rock is mined by open-cast mining almost everywhere in the world. Large mobile excavators remove the disintegrated rock in large blocks that are crushed and transported to storage. The crushed rock is then ground and, by means of a magnetic separator, the magnetite separated from the phosphate ore is obtained.

This ore is largely concentrated by the flotation process, resulting in an impure concentrate of apatite in the aqueous phase, which must pass through a thickener and a dryer to finally obtain a phosphate rock concentrate or phosphate concentrate with a 35 per cent P2O5 content, the term used commercially (VILLAVERDE, 2008).

The beneficiated rock, which is used as a raw material for the production of phosphoric acid, is generally insoluble, requiring a drastic chemical attack to obtain phosphoric acid. There are two known processes for producing phosphoric acid from phosphate rock: the dry process and the wet process.

In the dry process, the phosphate rock, together with coke and sand, is treated in an electric furnace at 1,300°C to release elemental phosphorus in the form of steam, which is condensed and then oxidised to P2O5, which, with water, produces high purity phosphoric acid, specifically for the pharmaceutical and food sectors. The disadvantage of this process is its high energy consumption (SANTOS, 2002).

The wet process is used in more than 90% of the world's phosphoric acid production areas (RUTHERFORD, 1994). This process is most commonly used to obtain phosphoric acid for the fertiliser industry and industrial phosphate for other industrial sectors, where the presence of impurities can be accepted at higher levels than in the pharmaceutical and food industries. The average energy consumption in this process is five times less than in the dry process (SANTOS, 2002).

The main Brazilian companies producing phosphate fertilisers are FOSFÉRTIL/ULTRAFÉRTIL (now Vale Fertilizantes), BUNGE do Brasil and COPEBRÁS, which together account for 96% of the total phosphate rock concentrate produced. Production takes place mainly in the states of Minas Gerais, Goiás and São Paulo (DNPM, 2008).

Figure 2.1 shows a simplified flowchart of the industrial process for obtaining phosphoric acid and the point at which phosphogypsum is generated (CEKINSKI, 1990).

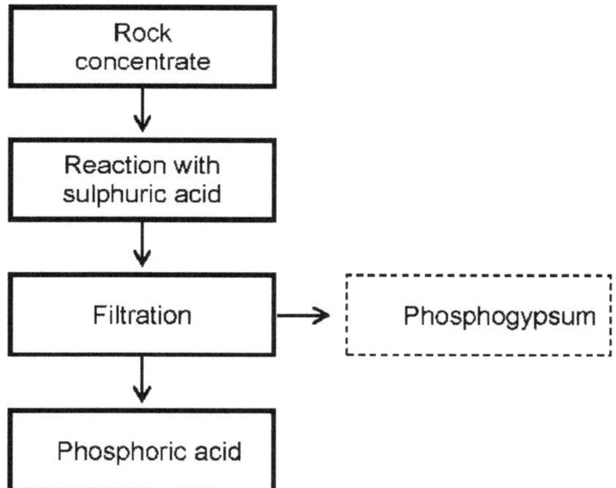

FIGURE 2-1 - SIMPLIFIED FLOWCHART OF THE INDUSTRIAL PROCESS FOR OBTAINING PHOSPHORIC ACID.

2.1.1.3 Physical and chemical characteristics

Phosphogypsum is a chemical compound with physical and chemical properties similar to those of gypsum. In differential thermal analysis (DTA) and thermogravimetric analysis (TGA), phosphogypsum showed a loss of free water at 96.4 ^{P}C, loss of part of the structural water at 155.4°C and a mass loss of approximately 33%.

The Hydrogen Potential (pH) of phosphogypsum ranges from 1 to 1.5 (CANUT, 2005). Phosphogypsum is acidic due to the residue of phosphoric acid, sulphuric acid and hydrofluoric acid contained in its pores, but it can reach neutrality if it is washed or leached with water. Phosphogypsum contains many impurities, such as quartz, fluorides, phosphates, organic matter, aluminium and iron minerals (RUTHERFORD, 1994).

Phosphogypsum has similar physical properties to natural gypsum, with a particle density of between 2.27 and 2.40 g.cm^3 (SENES, 1987).

According to CANUT (2005), the granulometric analyses of phosphogypsum show retention diameters of: 10% (D 10%) 5.819 pm, 50% (D 50%) 30.652 pm and 90% (D 90%) 65.005 pm.

The chemical and mineralogical characteristics of phosphogypsum depend on the nature of the phosphate ore, the type of process used, the efficiency of the plant operation, the age of the pile and some contaminants that may be introduced into the phosphogypsum in the production area (ARMAN, SEALS; 1990).

The phosphogypsum subjected to X-ray fluorescence chemical analysis shows the highest element content: sulphur (S) and oxygen (O), as shown in Table 2.1, with mineralogical analysis using X-ray diffraction illustrated in Figure 2.2.

TABLE 2-1 - CHEMICAL ANALYSIS OF PHOSPHOGYPSUM.

Sample	Higher content elements	Elements with medium content	Low element content	Dash
Phosphogypsum	S, O	Ca	P	Si, Al, Mg, Na, Fe, Ce, Ti, La, K, Sr, Zr, Pr

Source: CANUT (2005).

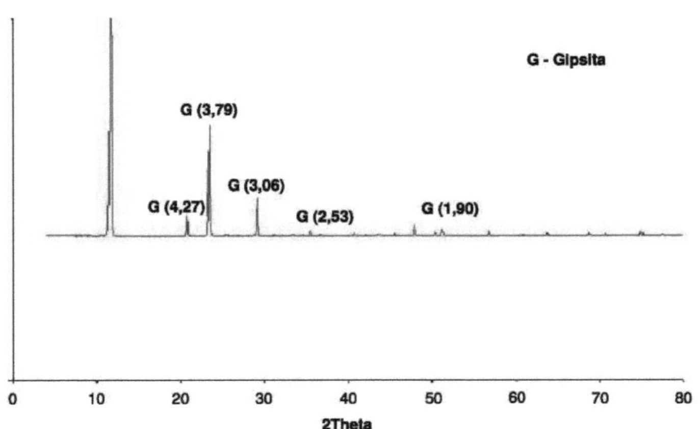

FIGURE 2-2 - X-RAY DIFFRACTION OF PHOSPHOGYPSUM. Source: CANUT (2005).

The conditions of the phosphoric acid production process have an influence on the morphology of the phosphogypsum crystal, which explains why crystals of different shapes are produced. The type of phosphate rock, the particle size of the phosphate rock, the concentration of the phosphoric acid, the solids content in the gypsum paste, the excess sulphuric acid in the gypsum paste, the impurities in the phosphate rock, the temperature and the reaction of the system (flow rate, feed volume, agitation and recirculation) are all factors that control the shape and size of the phosphogypsum crystals.

phosphogypsum (BECKER, 1989). Canut (2005) also observed the morphology of phosphogypsum and gypsum using the Scanning Electron Microscopy (SEM) method for comparison purposes, concluding that the micro-images are similar in their crystalline structure, as shown in Figure 2.3.

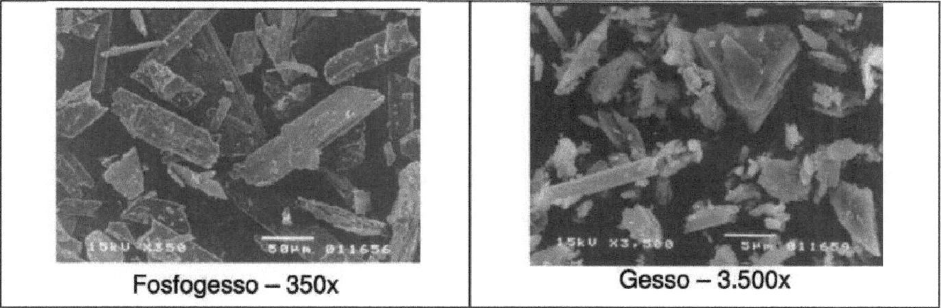

FIGURE 2-3 - MICROIMAGE OF GYPSUM AND PHOSPHOGYPSUM. Source: CANUT (2005).

Gypsum and phosphogypsum samples are made up of crystals in the form of interlocking rods and needles, formed during the hydration process of bassanite and are responsible for the cohesion of the mass (CANUT, 2005).

2.1.1.4 Availability of phosphate ores

The world's reserves of phosphate rock are approximately 50 billion tonnes, concentrated mainly in three countries: Morocco with 21 billion tonnes, which represents 42.3%, China with 13 billion tonnes, which represents 26% and the United States with 3.4 billion tonnes, which represents 7%, totalling 75% of the reserves. The remaining percentage is distributed mainly between the Republic of South Africa, Jordan, Australia, Tunisia, Russia, Syria, Egypt and Brazil (DNPM, 2008).

Brazil has phosphate rock reserves of 317 million tonnes, which are mainly located in the states of Minas Gerais (68%), Goiás (14%) and São Paulo (6%) (DNPM, 2008).

Most of Brazil's phosphate rock reserves, around 80 per cent, are of igneous origin, also known as magmatic, and are concentrated mainly in the states of Minas Gerais, São Paulo and Goiás (DNPM, 2008).

2.1.1.5 Phosphogypsum applications

The large production of phosphogypsum in recent years by the fertiliser industries in Brazil and the need for ever larger areas for its storage have boosted interest in recycling this industrial waste. In Brazil, to date, there are no specific rules regulating the use of phosphogypsum, so most of this waste is stored in the open, at the production site itself, and only a small part is reused (VILLAVERDE, 2008).

The main applications of phosphogypsum are in agriculture, cement production and construction, which is the subject of this study.

Application in agriculture

Agriculture is the largest domestic consumer of phosphogypsum (AQUINO, 2005). In agriculture, phosphogypsum is called agricultural gypsum and is commonly used as a source of nutrients for plants and in various physical and chemical processes in the soil profile, such as conditioning and improving sodic soils and clay soils. In addition, when applied to the soil surface, it is dissolved by rain and helps water infiltration, preventing surface hardening and favouring seed germination (NUERNBERG *etal*, 2005).

Phosphogypsum is a great source of nutrients such as calcium and sulphur. Calcium plays a fundamental role as a conditioner of the root environment, the effects of which are observed both in root growth and in the post-harvest resistance of fruit during storage. Sulphur is responsible for vital functions in plants, as it is a component of amino acids and takes part in metabolism (NUERNBERG *etal*, 2005).

Application in cement production

The commercial use of phosphogypsum or chemical gypsum in Brazil is mainly related to cement production. Consumption of phosphogypsum by the national cement industry amounts to 1.7 million tonnes per year (AQUINO, 2005).

To be used in the cement industry, phosphogypsum needs an acidity index of more than 4 and a phosphoric acid content of less than 0.8%. A low pH can contribute to the depassivation of reinforcement in reinforced concrete structures and reduce the initial strength of concrete. A high phosphoric acid content contributes to an increase in the setting time of cements (CANUT, 2006).

The use of phosphogypsum as an additive to clinker in the production of Portland cement, replacing natural gypsum, is an interesting alternative. In this case, calcium sulphate acts as an agent to control the setting time of the cement and is added at a content of up to 5%. Although viable, the use of phosphogypsum depends on the location of the fertiliser industries in relation to the cement plants, due to transport costs, and only 10% of the total quantity available in Brazil is used in this application (MAZZILLI *etal*, 2000).

Application in construction

Due to its similar physical and chemical properties to natural gypsum (gypsum), phosphogypsum can be used as a substitute in the manufacture of lining boards, panels, partitions, pre-moulded blocks, flooring and cladding. Natural gypsum has been used as a building material for millennia, since the construction of the pyramids in Egypt (HICAPIE *etal*, 1997).

In several countries, such as Japan, due to the scarcity of the raw material gypsum, phosphogypsum is widely used as natural plaster in construction (CANUT, 2006).

Gypsum is used in the construction industry as a fast-setting inorganic binder. This property allows the production of components such as blocks and slabs without treatment to accelerate hardening. Due to the specific properties of gypsum, these components have thermal and acoustic insulation and fire protection characteristics. In addition, gypsum paste can be used in finishes with excellent results (VILLAVERDE, 2008).

It should be borne in mind that the construction sector is recognised as one of those responsible for environmental impacts. This is mainly due to the large amount of exploited natural resources that are transformed and processed into material to be used in construction, such as natural gypsum reserves. There has therefore been a great deal of interest in replacing this material with phosphogypsum in the construction industry, both as a sealing material, for example prefabricated phosphogypsum-based boards, and as a finishing or internal cladding material in the construction of large-scale low-income housing. This practice could significantly reduce the cost of construction due to the wide availability of the material, benefiting a large part of the poorer population who live in more precarious situations (VILLAVERDE, 2008).

2.1.1.6 Environmental impacts of phosphogypsum

Phosphogypsum can be disposed of in two ways: in piles near fertiliser production plants or by pumping it into lagoons. Disposal in areas close to industries is the most common way of disposing of phosphogypsum and is often adopted almost everywhere in the world. In this situation there are two alternatives, "wet" (Figure 2.4 - A) or "dry" (Figure 2.4 - B). The final disposal is carried out "wet", where the phosphogypsum is disposed of together with the effluent from the industrial unit, in the form of slurry, by pumping it into sedimentation ponds, where it is decanted, and after drying it is accumulated in piles in the open air in areas specially designated for this purpose (VILLAVERDE, 2008).

The main routes of environmental contamination resulting from this storage are atmospheric contamination by fluorides and other toxic elements, pollution of groundwater by labile anions, acidity, trace elements and radionuclides, exhalation of radon, inhalation of dust and direct exposure to gamma radiation (VILLAVERDE, 2008).

In addition to the direct impacts linked to the generation of phosphogypsum and the final disposal of phosphogypsum from the phosphoric acid manufacturing industries, after the reuse of phosphogypsum in construction, there will be, at the end of its cycle, the generation of construction waste, together with rubble, and it must be stored, transported and destined in accordance with specific technical standards to the appropriate final destination in order to avoid an environmental impact (CONAMA, 2002).

Figure 2.4 shows the storage of phosphogypsum at a Brazilian fertiliser company (VILLAVERDE, 2008).

(A)

(B)

FIGURE 2-4 - "WET" (A) AND "DRY" (B) STORAGE OF PHOSPHOGYPSUM.

2.2 Gold Mining Sand

The installation of a mineral development usually provides the community located in its area of influence with an increase in jobs and income, the availability of goods and services, tax collection and an improved quality of life. On the other hand, it can also mean undesirable changes to the landscape and environmental conditions.

In gold extraction, as in other mining activities, the topsoil is removed which, after the extraction process, generates, among other materials, gold extraction sand (BORMA, SOARES; 2002).

2.2.1 Gold mining

Almost all the literature specialising in gold mining reports that this activity generally takes place in two ways: garimpagem (informal sector) and industrial mining. This classification is more common in technology publications, but it is also used by those focussing on environmental issues.

The technique of exploiting primary deposits in galleries can be observed in both prospecting and so-called industrial mining, in which case the use of explosives is necessary. However, in practice it is often difficult to discern

the two activities, especially when it comes to a large-scale business activity. Many mining companies use both forms of production on the same site, and may operate for a period with both techniques combined, or even successively, depending on the history of local exploitation.

Mining can take place on dry land or in the beds of watercourses. On dry land, the banks and slopes (the baixões) are usually dismantled with strong jets of water, but the mined ore is also fractionated using hammer mills and centrifuges. The material resulting from washing with a jet nozzle is dredged and conveyed to a wooden box using a "chupadeira". The box, which is predominantly long, is lined with a burlap sack or carpet and has transverse grooves. Mercury is placed at the top of the box and next to the slats so that it forms an amalgam with the gold particles present. Some of the mercury that is not combined with the gold is lost to the environment, as is the part that is amalgamated during the process of burning this alloy to purify the gold.

Gold extraction using the leaching technique with the use of cyanide or cyanide compounds is based on the phenomenon of percolation. The ore is extracted in its "raw" form (in combination with other elements and/or substances) from the deposit and is taken to be processed through leaching (CAHETÉ, 1995).

2.1.2 Gold mining in Brazil

Despite the growing exploitation of gold in Brazil, this production is still concentrated in a relatively small number of mines (PORTO *et al,* 2002). Data collected by THORMAN *et al* (2001) shows that between 1982 and 1999 more than 90% of mine production was concentrated in 17 main mines.

Gold production in Brazil has taken place over the last 20 years mainly in five regions: the Iron Quadrangle in Minas Gerais, the Carajás region in southern Pará, Itapicuru and Jacobina in Bahia and Crixás in Goiás (PORTO *etal.,* 2002).

2.1.3 Waste generated

The solid waste generated in mining and mineral processing operations can be categorised into tailings and waste rock.

The piles of this waste are generally very granular and, in the absence of compaction, have high porosity, which facilitates the penetration of gaseous oxygen and rainwater into their interior (BORMA, SOARES; 2002).

Among the sterile waste, one of the materials that surrounds gold is sand. As well as being used to rebuild the roof after extraction, sand can also be used in construction.

2.1.4 Sand for construction

The physical and chemical properties of aggregates and binding mixtures are essential for the life of the structures (works) in which they are used. There are countless examples of structures failing in which it is possible to conclude that the cause was the improper selection and use of aggregates (VALVERDE, 2001).

In nature, sand can be found in river sand harbours - which are the best - or in mines, when it is called "pit sand" or "ravine sand". These are the cheapest, but can contain impurities that need to be washed out before they can be used in more responsible work (CAMPOS, 2011).

In terms of type, sands are divided into coarse, medium and fine:

- Coarse sand - grains between 2 and 4 mm in diameter;
- Medium sand - grains with a diameter between 0.42 and 2 mm;
- Fine sand - grains with diameters between 0.05 and 0.42 mm.

2.1.5 Sand availability

Sand and rock for crushing are easily found in nature and are considered abundant mineral resources. However, this relative abundance must be treated with due care. Because they are products with a low unit value, the cost of transport makes the price more expensive for the end consumer.

Sand is extracted from riverbeds, floodplains, lake deposits, mantles of decomposed rock, pegmatites and decomposed sandstones. In Brazil, 90% of sand is produced in riverbeds. In the state of São Paulo, the ratio is different - 45 per cent of the sand produced comes from floodplains, 35 per cent from riverbeds and the rest from other sources. The main centres of sand production are the Paraíba do Sul River Valley, which accounts for 10% of national production, Sorocaba, Piracicaba and the Ribeira do Iguapé River Valley, all in São Paulo; Seropédica, Itaguaí, Barra do São João and Silva Jardim, in Rio de Janeiro; Guaíba, Caí and Jacuí rivers, in Rio Grande do Sul; Itajaí River Valley, in Santa Catarina; Iguaçu River Valley, in the Metropolitan Region of Curitiba, Tibagi River, in Ponta Grossa, and Paraná River, in Guairá, all in Paraná (VALVERDE, 2001).

2.3 Cal

Due to the multiplicity of its applications, lime - virgin and hydrated - is among the top ten mineral products consumed worldwide. The product gains even greater importance when you realise the wide range of industrial and social sectors that use it, thanks to its dual capacity as a chemical reagent and binder (CUNHA, 2007).

The knowledge of calcium as a chemical element is relatively recent. However, some of its compounds have been known since ancient times, as the Romans already used slaked lime (CaOH - calcium hydroxide), hydraulic lime and quicklime (CaO - calcium oxide) in construction (CUNHA, 2007).

Lime comprises six products resulting from the calcination of magnesian limestones/dolomites - hydrated or not. These products are calcium virgin lime (calcium oxide - CaO), calcium hydrated lime (calcium hydroxide - Ca(OH)2, normal hydrated virgin lime type N (calcium hydroxide, magnesium hydroxide and magnesium oxide - $Ca(OH)2.Mg(OH)2.MgO$, special hydrated dolomitic lime type S (calcium hydroxide and magnesium hydroxide - $Ca(OH)2.Mg(OH)2$, used in mortars, the intermediate types (classified as magnesian lime) of calcium and dolomitic lime is hydraulic lime used in structures (PETRUCCI, 1976).

2.3.1 Availability

Despite the considerable level of production - between 5 and 6 million tonnes per year in Brazil - lime has a low per capita consumption (approximately 36 $kg.year^{-1)}$. Even so, this rate is above the world average consumption of around 25 $kg.year^{1}$ (CUNHA, 2007).

In Brazil, the diversified areas of lime consumption are supplied by more than 200 producers spread across the country. The capacity of their facilities varies from 1 to 1,000 tonnes of quicklime per day. Brazilian lime production grew in 2007, increasing its share that year to 2.7 per cent of world production, ranking it fifth among lime-producing countries. Most of the lime produced in Brazil results from the calcination of metamorphic limestones/dolomites of different geological ages; generally very old (Precambrian) and of varying purity (PEREIRA, 2009).

China leads the world in lime production, with a share of more than 70 per cent, followed by the United States, which accounts for more than 8 per cent of this market, and Japan and Russia with almost 4 per cent each (PEREIRA, 2009).

2.3.2 Lime Production

The main product of the calcination of calcium and calcium-magnesian carbonate rocks is quicklime, also known as quicklime and ordinary lime. The term quicklime is used in Brazilian literature and in the standards of the ABNT - Brazilian Association of Technical Standards (2003) to designate the product composed predominantly of calcium oxide and magnesium oxide, resulting from the calcination, at a temperature of 900 - 1,200°C, of limestones, magnesian limestones and dolomites. The commercial quality of a lime depends on the chemical properties of the limestone and the quality of the firing (PEREIRA, 2009).

The process of making lime begins at the limestone quarry, where the limestone rock used to make lime is extracted. To extract the rock from the quarry, the process normally used is blasting with explosives.

The interest in limestone comes from its composition, which is basically made up of calcium carbonate [CaCOa] and magnesium carbonate [MgCOa], among others, compounds that will later be transformed into lime.

The stones are transported to the kiln, one of the main pieces of equipment in the process, where the limestone is calcined and transformed into lime.

For this important stage of the process there are various types of furnaces, ranging from vertical stone-built furnaces with manual or semi-automated processes, which usually have a low production capacity, to modern fully automated vertical or horizontal metal furnaces with a high production capacity (GARAY, 2011).

Lime is produced by calcining limestone and its reaction product is CO2, as shown in equation 2.2 (PATTON, 1978).

The main reaction that takes place inside the kiln during calcination is the decarbonation of the limestone, with 0 calcium carbonate and 0 magnesium carbonate being transformed into calcium oxide [CaO] and magnesium oxide [MgO], respectively (GARAY, 2011).

$$CaCO_3 + calor \rightarrow CaO + CO_2 \qquad (2.2)$$

After the calcination phase, which gives rise to quicklime, the hydration or extinguishing phase follows, which is the addition of water resulting in a very expansive and exothermic reaction, giving rise to slaked lime or hydrated lime (MARQUES, 2010). Hydrated or extinguished lime is produced by the reaction of quicklime with water, represented by equation 2.3:

$$CaO + H_2O \rightarrow Ca\,(OH)_2 \qquad (2.3)$$

The extinction reaction is accompanied by a great deal of heat production. Hydrated lime hardens in mortars thanks to its slow reaction with the carbon dioxide in the air (CO2), thus causing recarbonation, as shown in equation 2.4 (PATTON, 1978):

$$Ca\,(OH)_2 + CO_2 \rightarrow CaCO_2 + H_2O \qquad (2.4)$$

2.3.3 Lime applications

Despite being considered mankind's oldest manufactured product, new applications for lime are being discovered more and more frequently. In a theoretical inventory to show the composition of the world market, the ten most significant sectors can be identified, in descending order of quantitative expression (GUIMARÃES, 1985):

- The steel industry;
- The environment represented by the sub-sectors of wastewater treatment, acid mining water treatment, sulphur-rich fuel flue gas treatment and water treatment for drinking purposes;
- The pulp and paper industry;
- The alkali industry;
- The sugar industry;
- Non-ferrous mineralogy (copper, aluminium, gold, uranium and magnesium);
- The chemical industry (oil derivatives, tanning, grease, paints, calcium carbide and pharmaceutical products);
- Construction (mortars, building blocks and asphalt mixtures);
- Soil stabilisation;
- The ceramics industry (glass and refractories).

2.3.4 Lime production waste

In the years 2000 to 2002, it was common for cement plants to have disorganised waste storage areas that did not function properly, with waterproofed floors, roofs, contaminated rainwater treatment systems and areas that were properly demarcated and identified by type of waste. Currently, most facilities have waste warehouses that fulfil all the conditions described. A good practice in this area is the situation seen in two cement plants: *outsourcing* waste management, provided that the contract is made in accordance with clear rules and guarantees.

fully comply with legislation. Unlike the situation in the cement sector, in the hydraulic and non-hydraulic lime production sector, waste storage continues to be an area where conditions need to be improved. Lime production waste is badly burnt lime, which occurs when the combustion process has not taken place completely (GARCIA, 2008).

The use of alternative materials derived from industrial waste products in the manufacture of Portland cement is widely discussed. Waste such as limestone and lime-based carbonates are used by the cement industry in clinkerisation to conserve thermal energy and optimise production (BHATTYeta/, 2004).

2.4 Mortar

Mortar can be conceptualised as a complex material, made up essentially of inert materials of low granulometry (fine aggregates) and a paste with agglomerating properties, made up of minerals and water (active materials), which can also be made up of special products called additives (SABBATINI, 1986).

According to NBR 13529 (ABNT, 1995), mortar is a homogeneous mixture of fine aggregate(s), inorganic binder(s) and water, containing or not additives or additions, with adhesion and hardening properties. This same Brazilian standard defines other common terms involving coatings made from cement and lime, or both, in terms of their application. You can also find definitions such as:

- Lime mortar: mortar prepared with lime as the only binder;
- Simple mortar: mortar prepared with a single binder;
- Mixed mortar: mortar prepared with more than one binder.
- Additions: finely divided natural or industrial inorganic materials added to mortars to modify their properties and whose quantity is taken into account in the proportioning;
- Additive: a product added to mortar in small quantities in order to improve one or more properties, in the fresh or hardened state.

Applications are directly linked to the amount of binders used, the grain size of the sand and the amount of water added. In the construction industry, mortars are used for laying masonry, covering masonry (plastering, rendering and rendering), covering floors (screeds), laying various coverings (ceramics, stone, etc.), among others (RIBEIRO *etal*, 2002).

According to Scandolara (2010), mortar must have low cost, plasticity, adhesion, water retention, homogeneity, compactness, resistance to infiltration, traction and compression and durability. Each type of application requires different characteristics and properties that can be improved by including minerals and chemical additives.

Aggregates can be defined as fine aggregate or light aggregate. Binders are hydrated lime, quicklime, masonry cement, Portland cement and white Portland cement. Additions are: recycled rubble, ceramic phyllite, pozzolanic material, limestone powder, gravel, fine soil and processed soil. Additives can be considered a water repellent, air incorporator, permeability reducer and/or water repellent (ABNT, 1995).

2.4.1 Aggregates

Aggregates are produced from the crushing of rock masses (crushed stone, stone dust) or from the exploitation of natural particulate material (sand, pebbles or boulders).

The main application of aggregates is in the manufacture of concrete and mortar where, together with a binder (portland cement paste / water), they form an artificial rock with various uses in construction engineering, the main application of which is to make up the various structural elements of reinforced concrete (slabs, beams, pillars, footings, etc).

In the manufacture of concrete, the main purpose of using aggregates, whether sand or stone, is economic, as they are low unit cost materials, lower than cement. However, aggregates enable some other properties of the artificial rock to be formed to perform better, such as: reduced shrinkage of the cement paste, increased resistance to wear, better workability and increased resistance to fire (ARAÚJO *etal*, 2000).

2.4.1.1 Dimensions

In terms of size, aggregates are classified into two groups. Fine aggregates: quartz sands, coarse aggregates: pebbles, gravel, crushed stone and stone aggregates with large grains, as set out in the specifications of NBR 7211 (ABNT, 2005) and NBR 9935 (ABNT, 2011).

According to Brazilian specifications, light inorganic aggregates are divided into two groups: group I, those whose grains pass at least 98% of the 4.8 mm sieve. Group II is made up of coarse aggregates whose grains pass at least 90 per cent of the 12.5 mm sieve, according to the limits established in NBR 7213 (ABNT, 1984).

Light, medium or high density fine aggregates are: sand of natural or artificial origin resulting from the crushing and grinding of expanded vermiculite, stable rocks, barium ore and others or a mixture of all of them, whose grains pass at least 95 per cent of the 4.8 mm sieve according to NBR 2395 (ABNT, 1995).

According to NBR 7211 (ABNT, 2005), fine aggregates are "aggregates whose grains pass through the sieve with a mesh opening of 4.75 mm and are retained on the sieve with a mesh opening of 150 pm.

2.4.1.2 Fine aggregates

Sand, used as a fine aggregate in mortar and concrete, can be classified as natural (rivers, mines, floodplains) or artificial (fine quarry waste - stone dust).

Sand is extracted in mining units called sand pits or sand harbours, and can be extracted from river beds, lake deposits, underground sand veins (mines) or dunes. Most of the sand produced in Brazil comes from riverbeds or mined in pits flooded by groundwater. The sand, together with the water, is pumped into suspended silos or accumulated on the ground and then loaded onto tipper lorries bound for the distributor or the end consumer (ARAÚJO *et al,* 2000).

According to MINEROPAR (2004), aggregate has three main functions:

- Provide the binder with a relatively economical filler;
- Provide the paste with particles adapted to resist applied loads, mechanical wear and weathering percolation;
- Reduce volume variations resulting from the setting process, hardening and humidity variations in binder and water pastes.

Sand does not take part in the chemical reactions of mortar hardening (CARNEIRO *et al,* 1999). The particle size distribution of sand in the fresh state directly influences the performance of the mortar, affecting workability and water consumption, while in the finished coating it influences cracking, roughness, permeability and bond strength (ANGELIN *et al*, 2003).

2.4.2 Binders

Agglomerates, according to ARAÚJO *et al* (2000), are the active, binding material, usually powdery, whose main function is to form a paste that promotes bonding between the grains of the aggregate.

I. Airborne: these are binders that harden by the chemical action of CO2 in the air, such as airborne lime;
II. hydraulic: these are binders that harden by the exclusive action of water, such as hydraulic lime, Portland cement, etc. This phenomenon is called hydration;

III. Polymeric: these are binders that react due to the polymerisation of a matrix.

Usually portland cement (cement mortars), lime (lime mortars) and gypsum (gypsum mortars) are used as binders on construction sites.

2.4.2.1 Cement mortar

Portland cement mortar is essentially composed of cement, fine aggregate and water. It acquires high mechanical resistance in a short time, but has poor workability and low water retention. This type of mortar is used specifically for certain situations, such as making floors as a reinforced mortar, and is rarely used in masonry cladding. It is widely used to make plaster to be applied to masonry walls and concrete structures to increase the adhesion strength of mixed mortar coatings (SILVA, 2006).

2.4.2.2 Lime mortar

It consists of lime, fine aggregate and water. The lime paste fills the voids between the grains of the fine aggregate, improving plasticity and water retention (SILVA, 2006).

Lime can be used as the sole binder in mortars for laying bricks or cladding masonry or in mixtures to obtain soil/lime blocks, sand/lime blocks and alternative cements. Because of the high fineness of its grains (2 pm in diameter), and consequent ability to provide fluidity, cohesion (less susceptibility to cracking) and water retention, lime improves the quality of mortars. Lime gives pastes and mortars greater plasticity, allowing them to deform more without cracking than they would with Portland cement alone. Cement mortars containing lime retain more mixing water and thus allow for better adhesion (ARAÚJO *et al,* 2000).

In lime mortars, the presence of sand, as well as offering the advantages mentioned above, also facilitates the passage of carbon dioxide from the air, which produces the recarbonisation of calcium hydroxide (SHICHIERI, 2008).

2.4.2.3 Gypsum mortar

Of the binders used in construction, gypsum is the least used in Brazil. However, it has very interesting characteristics and properties, including rapid hardening, which allows components to be produced without hardening acceleration treatment. The plasticity of the fresh paste and the smoothness of the hardened surface are other important properties. Due to its main characteristic, rapid hardening, gypsum lends itself to moulding. Its main applications include: cladding material (stucco); boards for lowering ceilings (lining); panels for partitions and ornamental elements such as crown mouldings and flower boxes (ARAÚJO *et al,* 2000).

2.5 State of the art

Singh *et al* (1993) analysed the possibility of removing impurities from phosphogypsum before it is used in Indian cement industries, so that its impurities do not affect the strength of the cement. Aqueous hydroxide solutions were used to purify the phosphogypsum to make it suitable for cement manufacture. The results of the chemical and physical tests and differential thermal analysis of the phosphogypsum with and without the ammonium hydroxide treatment confirmed the effectiveness of the treatment for purifying phosphogypsum. The treated phosphogypsum had comparatively lower quantities of impurities than the untreated material, but the cement produced had mechanical properties similar to those produced using natural gypsum.

Singh (2002) also assessed the performance of phosphogypsum, after purification using an aqueous solution of citric acid, for use in the manufacture of cement and gypsum. The results of the chemical and physical tests and differential thermal analysis of phosphogypsum, with and without treatment with citric acid, show that Portland cements produced with purified phosphogypsum have physical properties similar to those produced from mineral gypsum and comply with Indian standards.

Yang *et al* (2009) studied the use of autoclaved phosphogypsum in the manufacture of load-bearing wall bricks. The crystalline phase, morphology and thermal characteristics of the original phosphogypsum waste were investigated. Next, Chinese standard size bricks were prepared from different compositions of phosphogypsum and autoclaved phosphogypsum. The results showed that the compressive strength of autoclaved phosphogypsum bricks was higher. The flexural strength and compressive strength of the bricks reached 4.0 MPa and 15.0 MPa, respectively. In conclusion, the use of autoclaved phosphogypsum to make wall load-bearing bricks was recommended instead of conventional burnt clay bricks.

Kuryatnyk *et al* (2008) presented the valorisation of phosphogypsum in conjunction with calcium sulfoaluminate clinker in the following proportions: 70% phosphogypsum and 30% calcium sulfoaluminate clinker. The tests lead to the production of a hydraulic binder, which means that it is not destroyed when immersed in water.

Drago *et al* (2009) analysed the technical feasibility of using basalt sand as a partial substitute for conventional sand in the production of concrete used in construction. They concluded that strength decreases as the percentage of artificial sand is increased, around 20% for total replacement. The results of the uniaxial compressive strength test showed a maximum value of 29.10 MPa at 28 days, using a composition of 50% mortar and 50% natural sand.

Corrêa and Mymrine (2005) sought to develop the best compositions for making composites based on concrete and lime production waste. They concluded that the increase in strength is directly proportional to the increase in the percentage of lime production waste in the compositions and the increase in curing time, and the compressive strength result showed a maximum value of 28.97 MPa at 28 days, using the composition of 60% concrete waste and 40% lime production waste.

Degirmenci (2007) researched the potential use of phosphogypsum with lime ash (lime production waste) in the construction industry. Phosphogypsum was used as a raw material and lime production waste to act as a binder. A series of tests were carried out to determine the compressive and flexural strength and water absorption after 28 days of sample preparation. Based on the results of the tests, it was concluded that the binder obtained can be used for the production of interior wall materials such as bricks and blocks.

Smadi *et al* (1999) carried out an experimental study to assess the potential use of phosphogypsum in concrete by preparing mixtures of mortar at a water/cement ratio of 0.6, using two types of cement, ordinary Portland cement and pozzolanic Portland cement, and two types of fine aggregate, natural river sand and limestone. These mixtures were prepared in different compositions of FG and purified FG ranging from 10 to 100 per cent. The purified FG was obtained by calcination at temperatures of 170°C, 600 °C, 750 °C, 850 °C and 950 °C. The compressive, tensile and flexural strengths of different hardened mortars were analysed, at a particular percentage of FG substitution, the results indicated a tendency for the strength properties to increase over the curing time. The purification process, by heating FG up to 900 °C, resulted in an improvement in the strengths of the mortar mix. The incorporation of GF into the cement paste drastically increased its initial and final setting times.

Sebbahi *et al* (1997) analysed the thermal behaviour of phosphogypsum from Morocco in the absence of additives. Thermogravimetric analysis (TGA), differential thermal analysis (DTA), infrared spectroscopy and X-ray diffraction tests were carried out. The DTA showed the transformations that had taken place. In addition, TGA showed that dehydration can occur in one phase, which confirms that the sample is not bihydrated. Infrared spectroscopy made it possible to characterise the water molecules from crystallisation and in the gypsum sulphate cluster.

In research related to the engineering properties of unpurified phosphogypsum in mixtures, Chang et al (1990) state that the results indicated that in order to utilise unpurified FG it is necessary to assess the source of origin before use. However, with adequate selection and engineering controls, FG has potential for use in road construction applications in the industry, including building and precast construction products such as bricks and masonry blocks.

In his research, Mazzili (2005) found that the radioactivity measured in phosphogypsum is of the same magnitude as that observed in fertilisers, which makes it feasible to reuse phosphogypsum as an input for the construction industry.

Research carried out in northern Vietnam to obtain new materials using industrial waste used lime production waste powder with low-quality portland cement to produce a cement composite, obtaining a high-performance material (STROEVEN *etal*, 2001).

CHAPTER 3

METHODOLOGY, RESEARCH METHODS AND MATERIALS USED

3.1 Research methodology

The following methodology was developed to meet the objectives of the study:

- Collection of representative samples of the chosen industrial waste by means of the quartering method;
- Characterisation of the physical, chemical and mineralogical composition of industrial waste to enable its reuse as components of new materials;
- Preparation of specimens with different compositions of the initial components;
- Study of the mechanical properties of CPs produced at different hydration and curing ages;
- Study of the physical-chemical processes of the chemical interactions between the components during the curing time, seeking to improve the final properties of the materials developed.

3.2 Physical, chemical and microscopic tests

The following tests were carried out to characterise the raw materials: X-ray fluorescence, particle size test, scanning electron microscopy, chemical micro-analysis using the energy dispersion method, uniaxial compressive strength, water resistance coefficient, water absorption by immersion, linear expansion coefficient and X-ray diffraction.

3.2.1 X-ray fluorescence

The X-ray fluorescence method was used to analyse the chemical composition of the raw materials, in an elementary and non-destructive way. The analysis was carried out on the Philips/Panalytical model PW 2400 equipment, with a 3 kW tube and rhodium target, at the UFPR Minerals and Rocks Laboratory (LAMIR).

3.2.2 Particle size test

The wet granulometric test, which consists of granulometric distribution by sieves, was carried out at UFPR's LAMIR, using four overlapping sieves with mesh sizes of 710 pm, 355 pm, 250 pm and 180 pm.

For sieving, 50 grams of sample were used, weighed on a balance with a precision of four places. After sieving, the fractions were measured by mass (g) to calculate their percentage.

Due to the fact that phosphogypsum particles have a lamellar shape, the laser particle size test is not the most suitable way to measure the diameter of these materials, as the measurement is carried out according to the position of the particles passing through the laser beam.

3.2.3 Scanning Electron Microscopy

Scanning Electron Microscopy (SEM) was used to investigate the morphologies of the raw materials and the changes in the structures of the materials developed. The microscope used was a JEOL model JSM-6360 LV at the Bosch Scanning Electron Microscopy Laboratory.

To cover the surface, the materials were metallised with gold in an SCD 030 - BALZERS UNION SL 9496 machine at the UFPR Electron Microscopy Centre.

3.2.4 Chemical micro-analysis by the Energy Dispersion method

Coupled with the SEM, the Energy Dispersive Spectrometer (EDS) carries out chemical micro-analysis to obtain the chemical elements found in the sample at a specific point or in a predetermined area by scanning electron microscopy.

3.2.5 Uniaxial compressive strength

The uniaxial compressive strength tests were carried out in accordance with NBR 12.129/91 on the EMIC DL 10.000 universal testing machine at the Materials and Surface Treatment Laboratory at UFPR. The load applied to the specimens until they broke was 500 N/s.

3.2.6 Water resistance coefficient

The water resistance coefficient (CA) was determined by the uniaxial compressive strength of the CPs on the 28th day, saturated after total immersion in water for 24 hours, and the water resistance of the CPs under ambient conditions, obtained using equation 3.1:

$$C_A = \frac{R_{SAT}}{R_{AMB}} \quad (3.1)$$

Where:

RSAT = uniaxial compressive strength of saturated test cups after total immersion in water for 24 hours

RAMB = uniaxial compressive strength of the specimens under ambient conditions.

3.2.7 Water absorption by immersion

To obtain immersion water absorption (Abscp), 6 PCs were used for each of the 12 compositions with a curing time of 28 days, in accordance with standard NBR 9778/2009, where the weight of the dry PC is compared with the weight of the saturated PC after immersion in water for 24 hours. Abscp is determined according to equation 3.2:

$$Abs_{Cp} = \frac{(M_{sat} - M_s)}{M_s} x100 \quad (3.2)$$

Where:

Msat = mass of saturated CP

M_s = mass of dry CP

1.1.8 Linear Dilation Coefficient

The Linear Dilation Coefficient (a) was used to check the dilation of the samples during hydration and curing. The initial diameter of the CPs was measured using a caliper and, after the curing times, it was remeasured to calculate the Linear Dilation Coefficient, according to equation 3.3:

$$\alpha = \frac{(l_F - l_I)}{l_I} x100 \quad (3.3)$$

Where:

h = initial diameter

IF = final diameter

All measurements were carried out in the Environmental Technology Laboratory at UFPR, under the same ambient conditions.

1.1.9 X-ray diffraction

In order to study the structures of the raw materials and the change in mineralogical composition during the curing of the materials developed, X-ray diffraction tests were carried out using the Philips PW 1.830 equipment, with a long-focus diffraction X-ray tube, at LAMIR in UFPR, which has the X'Pert recognition software and the JCPDS database as its analytical base.

3.3 Material Selection

The choice of waste was based on the principle of mortar, which is often used in construction, consisting of at least one binder, fine aggregates and water. Phosphogypsum and lime production waste were used as the binder and gold extraction sand as the fine aggregate.

3.3.1 Phosphogypsum

This study used a sample of phosphogypsum from a small Brazilian phosphorus fertiliser company in the state of São Paulo. This company has 42,000,000 tonnes of phosphogypsum disposed of in its own industrial landfill. Currently, 2,000,000 tonnes of its annual production is being used as a fertiliser or soil acidity neutraliser, but 2,000,000 tonnes.year^{-1} is still being disposed of in the factory landfill.

3.3.2 Gold Mining Sand

The sample of gold extraction sand used in this study was taken from a gold extraction mine located in the metropolitan region of Curitiba, Paraná. The gold content of this mine reaches 7 grams per tonne of ore. The data collected in this study was aimed at understanding the processes used in gold extraction and characterising the sand as a potential material for use in construction.

3.3.3 Lime Production Waste

Lime production waste collected from a lime manufacturing industry located in Rio Branco do Sul, Paraná, was used to make the composite. Data was collected on the production of lime and the applications of the waste generated in the production process.

3.4 Preparation of Test Specimens for the New Materials

The compositions chosen for this work were based on preliminary analyses, in which various compositions were prepared, varying the percentage of raw materials. Twelve compositions were selected from Table 3.1. The proportion of water used is 1:1 in relation to the lime production residue, which varied between 10 and 20 per cent.

After being weighed on a scale with a precision of four places, the materials were homogenised in a porcelain grater.

The samples, measuring 20 x 20 mm and weighing 13 grams (Figure 3.1), were defined according to the available mould.

TABLE 3-1 - COMPOSITIONS WORKED ON.

N°	Compositions (%)		
	Phosphogypsum	Gold Mining Sand	RPC
01	60	25	15
02	70	10	15
03	65	15	20
04	60	20	20
05	55	30	15
06	50	40	10
07	45	45	10
08	40	45	15
09	25	55	20
10	15	65	20
11	10	70	20
12	05	75	20

The specimens were made in cylindrical steel moulds, a non-absorbent material that is chemically inert to the waste being processed, with the following internal dimensions: 20 x 60 mm.

The CPs were compacted using a BOVENAU manual press with a capacity of up to 15 tonnes, with a pressure of 10 MPa and a pressure dwell time of 30 seconds.

FIGURE 3-1 - PHOTO OF A SPECIMEN MADE.

To obtain the coefficient of expansion, 6 specimens were made for each of the 12 compositions studied, for each curing time: 03 days, 07 days, 14 days, 28 days, 60 days, 90 days, 180 days, 12 months and 18 months. These specimens were used to carry out the uniaxial pressure resistance test, X-ray diffraction and Scanning Electron Microscopy on the fracture of the specimens.

For the water absorption tests and to obtain the water resistance coefficient, 6 specimens were prepared 28 days after curing, totalling 66 specimens per composition, as shown in Table 3.2.

TABLE 3-2 - NUMBER OF SPECIMENS PER COMPOSITION.

Healing Time Tests	3 days	7 days	14 days	28 days	60 days	90 days	180 days	1 year	1,5 years	Total
1° - Coefficient of expansion 2° - Uniaxial compressive strength	6	6	6	6	6	6	6	6	6	54
Water resistance coefficient	-	-	-	6	-	-	-	-	-	6
Water absorption by immersion	-	-	-	6	-	-	-	-	-	6
TOTAL	66 Specimens by composition									

CHAPTER 4

RESULTS AND DISCUSSIONS

4.1 Raw materials

Phosphogypsum, Gold Extraction Sand and Lime Production Waste (raw material) were characterised, assessing their chemical composition and specific structures.

4.1.1 Chemical composition of raw materials

From the semi-quantitative X-ray fluorescence analyses, an estimate of the chemical composition of the raw materials used was obtained, with results expressed in oxides, as shown in Table 4.1. Phosphogypsum showed a high concentration of Sulphur - SO3, 45%, and Calcium - CaO, 31%. The predominant element in the composition of gold extraction sand is 0 Silicon, SÍO2, 77%. Lime production waste has 0 Calcium - CaO, 48%, and 0 Magnesium - MgO, 33%, as its main constituents.

TABLE 4-1 - CHEMICAL COMPOSITION OF THE RAW MATERIALS USED.

Oxides	Chemical composition, mass %		
	Phosphogypsum	Gold mining sand	Lime production waste
Al_2O_3	0,20	9,80	0,25
BaO	0,50	0,20	-
CaO	30,70	1,40	47,69
Fe_2O3	0,20	1,20	0,24
K_2O	-	4,40	0,05
MgO	<0,1	0,70	33,10
MnO	-	<0,1	-
Na_2O	-	1,40	-
P205	0,90	0,30	0,02
SiO_2	0,90	76,90	2,82
SO_3	45,00	0,30	0,03
TiO_2	0,60	0,30	0,04
P.F.	18,93	2,37	15,74

4.1.2 Raw material granulometry

The particle size distribution of the three residues used as raw materials was determined by the particle size test, by determining the mass that each specified particle size range represents in the total mass tested. The results are shown in Table 4.2.

TABLE 4-2 - PARTICLE SIZE DISTRIBUTION OF RAW MATERIALS.

Opening (mm)	Cumulative retention (%)		
	Sand Gold Mining	Phosphogypsum	Lime Production Waste
0,710	39,68	0,22	0,12
0,355	13,97	28,23	5,31
0,250	10,13	12,48	10,53
0,180	8,22	13,71	16,30
Fine	28,00	45,35	67,74

Table 4.2 shows that AEO has 40 per cent particles larger than 0.71 mm and 28 per cent particles smaller than 0.18 mm. Particles with a particle size of less than 0.18 mm prevail in almost half of FG's composition. In RPC, the finest particles prevail - 67.74%, which influences the increase in chemical activity of this binder when mixed with other components of the materials developed.

4.1.3 Raw material structures

4.1.3.1 X-ray diffractogram of phosphogypsum

The mineralogical structures were studied using X-ray diffractometry. Figure 4.1 shows the peaks of crystalline materials and the base of amorphous materials. The minerals in phosphogypsum are shown in Table 4.3.

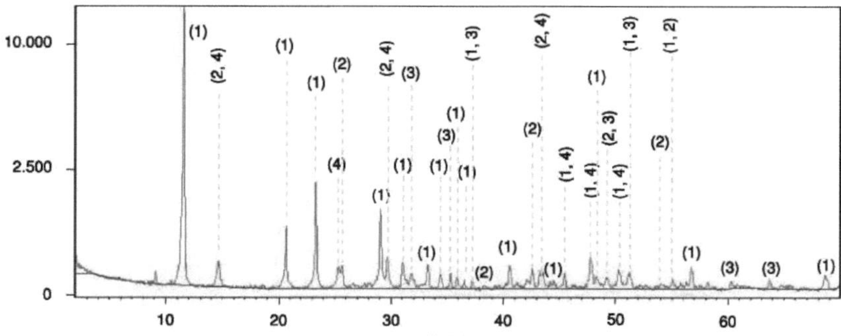

FIGURE 4-1 - PHOSPHOGYPSUM DIFFRACTOGRAM.

The phosphogypsum showed peaks of gypsum, hydrated calcium sulphate, dolomite and magnesium oxide carbonate. The dolomite probably comes from contamination of the phosphogypsum during storage.

TABLE 4-3 - MINERALOGICAL COMPOSITION OF PHOSPHOGYPSUM.

N°	Minerals	Formula
1	Plaster	$Ca\ SO_4\ 2H_2O$

2	Calcium Sulphate Hydrate	$CaSO_4\ 0.67\ H_2O$
3	Dolomite	$CaMg(CO_3)_2$
4	Magnesium Oxide Carbonate	$Mg_3O(CO_3)_2$

Determining the mineralogical composition of the raw material is fundamental to understanding the individual minerals and then comparing them with the transformation of the mineralogical composition of the new materials obtained.

Figure 4.1 shows the largest peaks of the gypsum mineral, which characterises the material as a binder.

4.1.3.2 SEM and EDS results of phosphogypsum

The morphology of the phosphogypsum samples, as a raw material, was studied using the Scanning Electron Microscopy (SEM) method, the results of which are shown in Figure 4.2.

The chemical composition was studied using the Energy Dispersive System (EDS) method, and the results obtained are shown in Table 4.4.

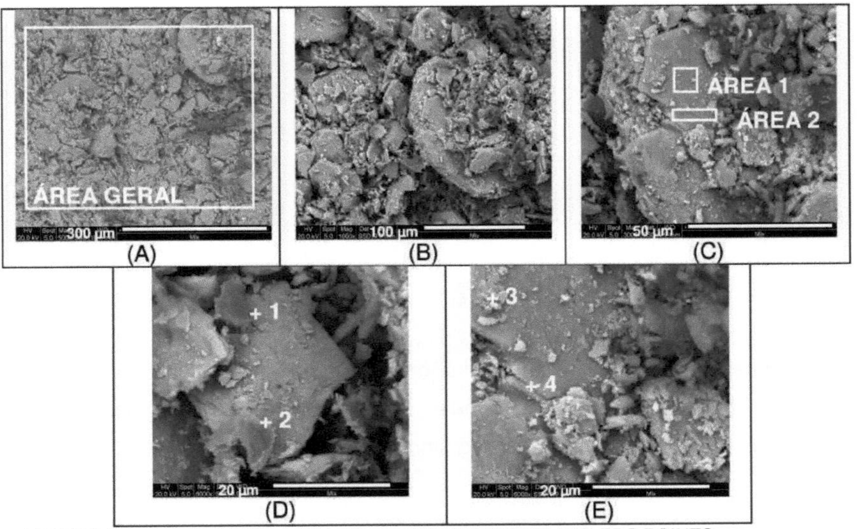

FIGURE 4-2 - MICROIMAGE (MEV) OF PHOSPHOGYPSUM AND EDS POINTS.

In the micro-images in Figure 4.2 it can be seen that all the phosphogypsum particles have different sizes and configurations, with no physical or chemical bonds between them. These characteristics are more visible especially with increased magnification, and different particle shapes can be seen: round, oval, angular, flat hexagonal (Figure 4.2 - C), laminar, needles, curved, among others. Pores of different sizes and configurations can be seen between the particles.

Areas 1 and 2 in Figure 4.2 - C have a similar chemical composition to the General area, with different but not significant values. In Figure 4.2

- C it is visible that areas 1 and 2 have crystals with hexagonal and laminar shapes typically found in natural gypsum crystals.

The general chemical composition of the surface of the sample studied (Figure 4.2 - A) shown in Table 4.4 shows that the majority of the composition is made up of Calcium (52.41%) and Sulphur (33.70%). Among its composition, Carbon (12.80%) is present as an element of natural carbonates. The general composition of the sample is similar to the chemical composition of natural gypsum ($CaSO_4$).

TABLE 4-4 - RESULTS OF THE EDS ANALYSIS - CHEMICAL COMPOSITION OF THE PHOSPHOGYPSUM AREAS AND POINTS.

Spectrum	Chemical composition (EDS) of the phosphogypsum areas and points, relative %				
	C	Si	S	Ca	Total
General	12,80	1,09	33,70	52,41	100,00
Area 1	17,30		21,86	60,84	100,00
Area 2	14,46		25,80	59,75	100,00
Point 1	17,21		37,56	45,13	100,00
Point 2	16,14		38,81	45,05	100,00
Point 3	25,26		33,77	40,97	100,00
Point 4	19,81		37,67	42,52	100,00

Points 1, 2, 3 and 4 of the phosphogypsum also have a similar chemical composition to those mentioned above, with different Carbon, Sulphur and Calcium values, probably due to contamination with other materials.

4.1.3.3 X-ray diffractogram of gold extraction sand

The mineralogical structures of the gold mining sand were studied using X-ray diffractometry. Figure 4.3 shows the peaks of crystalline materials and the base with amorphous materials. The minerals in the gold mining sand are shown in Table 4.5.

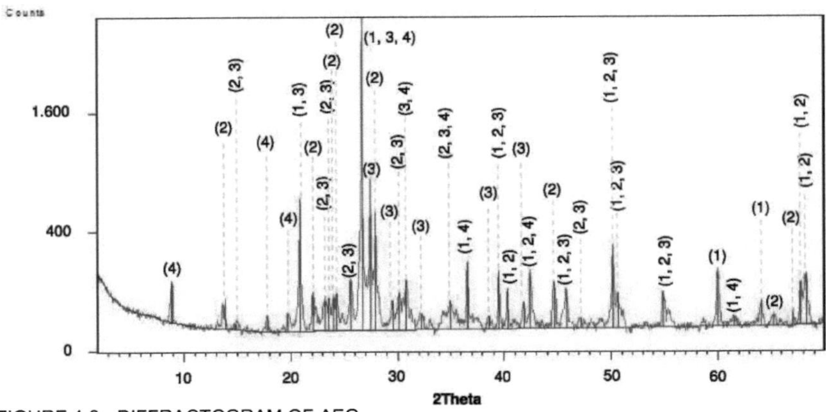

FIGURE 4-3 - DIFFRACTOGRAM OF AEG.

The gold mining sand showed peaks of quartz, albite, microcline and muscovite.

TABLE 4-5 - ESTIMATION OF THE MINERALOGICAL COMPOSITION OF GOLD MINING SAND.

N°	Minerals	Formula
1	Quartz	SiO_2
2	Albita	Na Al Si_3 O_8
3	Microcline	K Al Si_3 O_8
4	Muscovy	(K, Na) (Al, Mg, Fe) $_{(2)}$ ($Si_{3,1}$ $Al_{0,9}$) O_{10} (OH) $_{(2)}$

Figure 4.3 shows the largest peaks of the quartz mineral, which characterises the material as aggregate.

4.1.3.4 SEM and EDS results of AEO

Using the Scanning Electron Microscopy (SEM) method, the morphology of the Gold Extraction Sand samples used in this study as raw material was investigated; the micro-images are shown in Figure 4.4. The chemical composition was estimated using the EDS method and the results obtained are shown in Table 4.6.

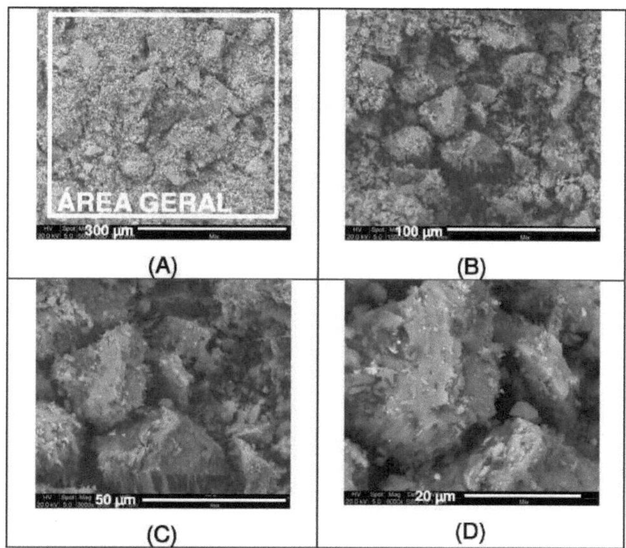

FIGURE 4-4 - MICROIMAGE (MEV) OF AEO AND EDS AREA.

In the micro-images (Figure 4.4) of the gold extraction sand, it can be seen that the particles have no bonds between them and have very irregular configurations and sizes. With increasing magnification, the absence of alloys is even more evident.

Microimages C and D (Figure 4.4) show wear on the surface of the particles and the absence of trigonal crystalline formations made up of silica tetrahedrons, normally characteristic of quartz particles, in contradiction to the XRD. The wear on the surface was possibly due to the influence of weathering by natural agents such as rain, wind, transport and sedimentation in the waters at different natural environments (alkaline and acidic) or even the influence of the technological processes involved during the gold extraction process.

Table 4.6 shows the results of the general chemical composition of the Gold Extraction Sand, which is mostly composed of Silicon (74.99%), in line with the results obtained in the X-ray fluorescence test. In lower concentrations are Potassium (14.09%), Aluminium (14.01%), Calcium (2.96%), Iron (2.06%), Sodium (1.17%) and Magnesium (0.70%). This composition is characteristic of natural sand.

TABLE 4-6 - RESULTS OF THE EDS ANALYSIS - CHEMICAL COMPOSITION OF THE AREAS AND POINTS OF THE AEO.

Spectrum	Chemical composition (EDS) of the gold mining sand area, relative %							
	In	Mg	Al	Yes	K	Ca	Fe	Total

General	1,17	0,70	14,01	74,99	14,09	2,96	2,06	100,00

4.1.3.5 X-ray diffractogram of lime production waste sand

X-ray diffractometry was used to study estimates of the mineralogical structures of the lime production waste. Figure 4.5 shows the peaks of crystalline materials and the base with amorphous materials. The minerals in the lime production waste are shown in Table 4.7.

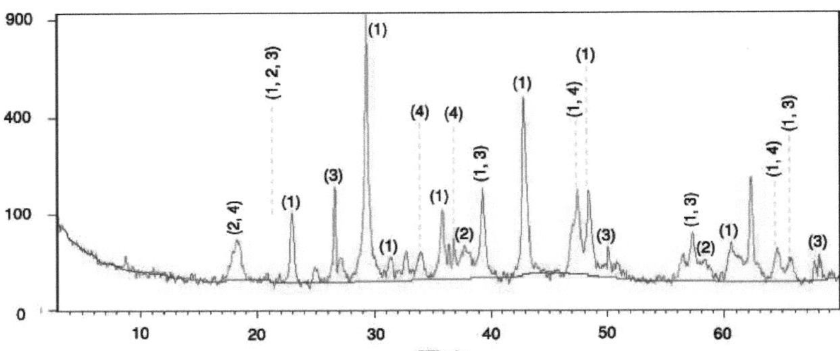

FIGURE 4-5 - DIFFRACTOGRAM OF LIME PRODUCTION WASTE.

The lime production waste showed peaks of Calcite, Brucite, Quartz and Portlandite. Calcite, which predominates in lime production waste, is found in carbonated lime before the hydration process.

TABLE 4-7 - ESTIMATION OF THE MINERALOGICAL COMPOSITION OF LIME PRODUCTION WASTE.

N°	Minerals	Formula
1	Calcite	Ca CO3
2	Brucita	Mg (OH) 2
3	Quartz	SiO2
4	Portlandite	Ca (OH) 2

4.1.3.6 SEM results of lime production residues

The morphology of the Lime Production Waste samples, as a raw material, was studied using the Scanning Electron Microscopy (SEM) method; the micro-images are shown in Figure 4.6.

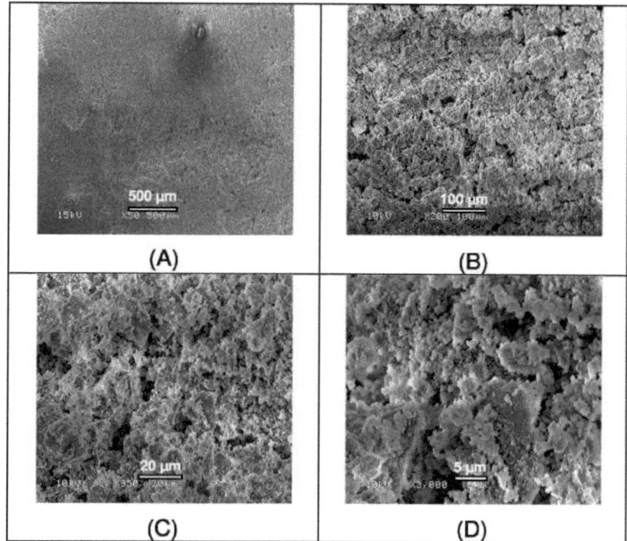

FIGURE 4-6 - MICRO-IMAGE OF LIME PRODUCTION WASTE.

Figure 4.6 - A shows the smooth, uniform surface, without elevations, with a network of irregular pores, increasing the contact surface of the material. In Figure 4.6 - B these pores are more visible and their shapes are irregular. The particles do not have very visible crystalline shapes, with asymmetrical edges.

The asymmetrical shapes are more visible in the enlargements of Figures 4.6 - C, and especially in Figure 4.6 - D, with formations that are similar to amorphous ones.

4.1.4 Mechanical properties

4.1.4.1 Uniaxial compressive strength

Table 4.8 shows the change in uniaxial strength up to 18 months of curing of all the compositions studied. The strength values ranged from 1.8 MPa (composition 8) on day 3° to 13.7 (composition 4) at 1 year of curing.

The highest results in terms of resistance to uniaxial compression occurred with a curing time of 1 year in composition 4, but the other compositions did not show the best resistance results at the same curing time, so it is not possible to state the ideal curing time for the material. It was observed that the variation in resistance depends on the composition and the curing time of the material; the longer the curing time, the higher the resistance results.

TABLE 4-8 - UNIAXIAL RESISTANCE RESULTS.

Nfi	Compositions, mass %			Uniaxial strength (MPa) after days/years of curing								
	FG	AEO	RPC	3 days	7 days	14 days	28 days	**60 days**	90 days	180 days	1 year	1.5 years
1	60	25	15	7,7	6,2	7	7,6	7,4	7,6	8,7	9,7	10,1
2	70	15	15	6,9	6,3	5,8	5,6	5,7	9,4	9,3	9,4	9,5
3	65	15	20	5,6	6,9	7,4	7,1	6,9	11,4	9,7	11,3	9,8
4	**60**	20	20	8,3	7,8	**9,9**	9,8	**11,6**	13,5	12,7	13,7	12,9
5	55	30	15	7,5	7,3	7,2	7,7	10	9,4	9,5	9,7	9,8
6	50	40	10	6,8	7,2	7,3	7,8	7,2	9	8,6	9,1	8,5
7	45	45	10	5,6	6,4	6,5	6,3	7	8,8	8,7	8,8	8,6
8	40	45	15	1,8	2,5	3,6	5,2	5,8	6,8	6,2	6,7	6
9	25	55	20	3,2	4,9	5,2	5,5	6,7	8,2	8,7	9	9,5
10	15	65	20	2,6	3,5	4	4,8	6,3	8,6	9,1	9,7	10,3
11	10	70	20	2,4	3,6	4,2	5,4	6,7	8,8	8,5	8,7	8,5
12	5	75	20	1,9	3,1	4,4	5	5,8	6,5	7,3	6,9	8,4

On the 3rd day of curing, only compound 4 complies with NBR 13.207/1994 - Gypsum for civil construction (Uniaxial Strength > 8.4 MPa at 28 days of curing), but at 90 days most of the compounds meet this demand. In addition, the strength values of the compositions also meet the requirements of NBR 7170/1983 for Solid Bricks: Class A < 2.5 MPa, Class B 2.5 < 4.0 MPa and Class C > 4.0 MPa, and for Ceramic Blocks: Class 15 = 1.5 < 2.5 MPa, Class 25 = 2.5 < 4.5 MPa on the 3rd day.

The compositions with a higher AEO content had lower uniaxial strength values, possibly due to a lower binder content.

Considering that sand is the inert material in the composition, the variation in uniaxial compressive strength for different levels of phosphogypsum and lime production waste after 28 days and 1.5 years of curing can be seen in Figures 4.7 and 4.8.

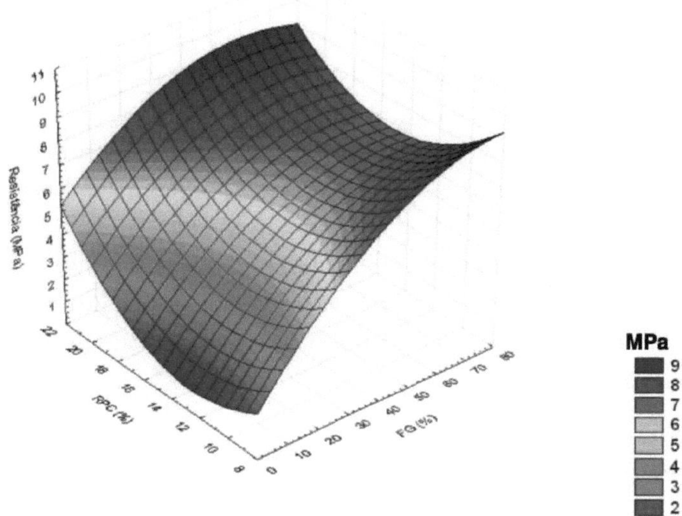

FIGURE 4-7 - VARIATION IN STRENGTH AT DIFFERENT FG AND RPC CONTENTS, AFTER 28 DAYS OF CURING.

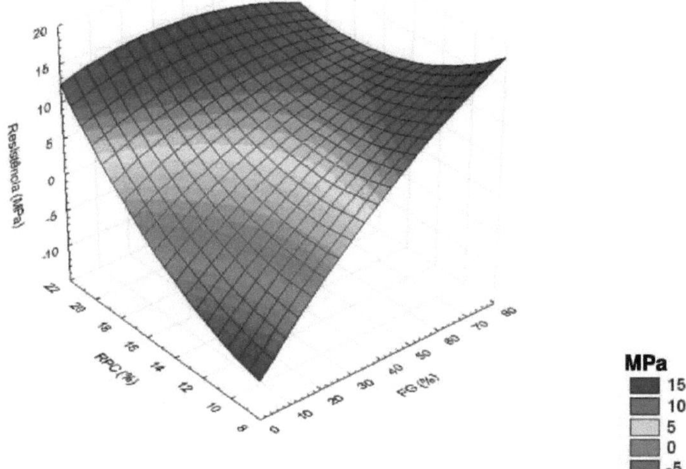

FIGURE 4-8 - VARIATION IN RESISTANCE AT DIFFERENT FG AND RPC CONTENTS AFTER 1.5 YEARS OF CURING.

At 28 days of curing, the highest compressive strength values were seen in compositions made up of between 45 and 75 per cent FG and between 20 and 22 per cent RPC, while the lowest compressive strength results were seen in compositions made up of between 6 and 15 per cent FG with RPC between 11 and 19 per cent.

After 90 days of curing, the highest compressive strength was found in the 70 to 80 per cent FG range, so the FG content has a strong relationship with the strength of the material. The lowest strength values correspond to the interval between 10 and 13 per cent RPC and 10 and 15 per cent FG. These values show lower binder contents and, consequently, higher aggregate contents, i.e. AEO.

Comparing compositions 3 and 4, with the same RPC content, the results show that with the addition of 5% phosphogypsum, composition 3 showed a reduction in strength results at all curing times.

Most of the compositions show a non-linear variation in strength with curing age, and the graph of the variation in strength with curing time in figure 4.9 shows an accelerated growth that stabilises around 90 days. Composition 4 showed the highest resistance results throughout the curing process studied.

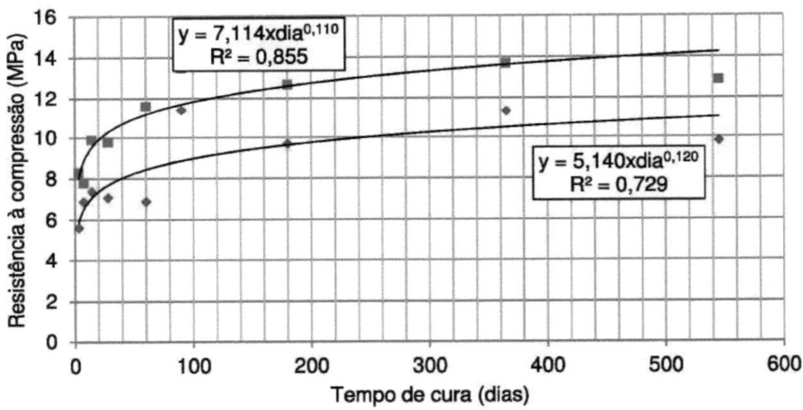

FIGURE 4-9 - VARIATION IN STRENGTH OF COMPOSITIONS 3 AND 4 WITH CURING TIME.

Compositions 3 and 4 show that the gradual increase in curing time results in a marked increase in compressive strength. This increase in strength, represented by the angular coefficient on the trend line, depends on the composition of the material studied.

The evolution of resistance during the 90 days of curing, when the sharp increase in resistance occurs, is shown in Figure 4.10. The curing time showed a linear relationship with the different curing times analysed.

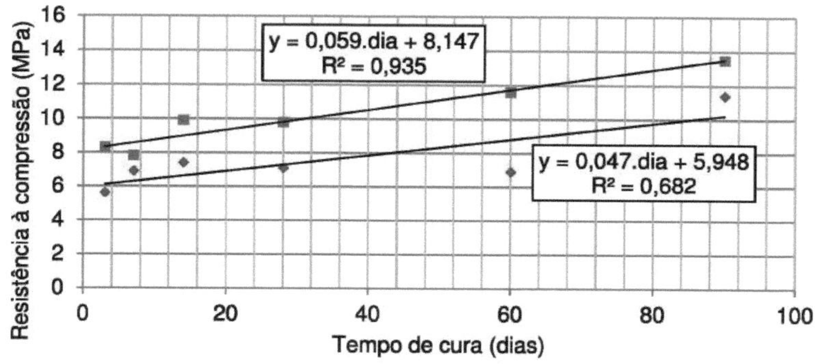

FIGURE 4-10 - VARIATION IN STRENGTH OF COMPOSITIONS 3 AND 4 DURING 90 DAYS OF CURING.

As shown in Figure 4.11, comparing compositions 09, 10, 11 and 12, with the same RPC content, it is clear that the AEO content helps to increase the strength values up to 65% (compositions 09 and 10), while higher AEO contents (compositions 11 and 12) cause a reduction in strength, as they reduce the compactness of the material. The increase in AEO, at the same RPC content, decreased the FG content, which possibly also contributed to the reduction in compressive strength.

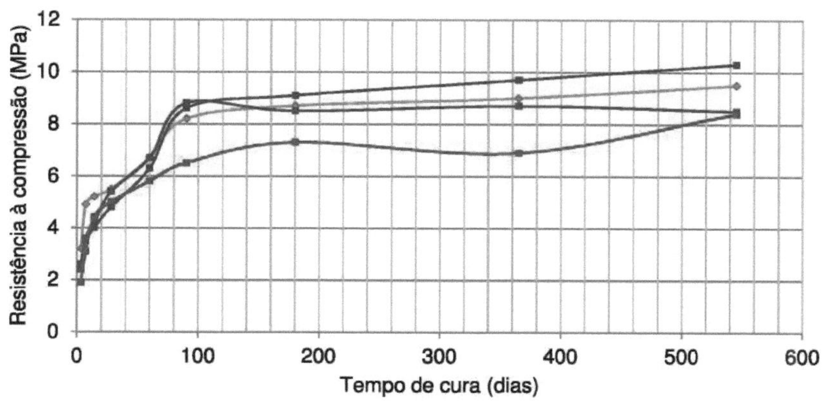

FIGURE 4-11 - VARIATION IN STRENGTH OF COMPOSITIONS 9, 10, 11 AND 12 WITH CURING TIME.

4.1.4.2 Water resistance

Water resistance tests (Table 4.9) were carried out on samples of all the compositions with a curing time of 28 days, with the aim of calculating the coefficient of water resistance. The results in Table 4.9 show that all the compositions, with CA ranging from 0.76 to 0.95, presented satisfactory results, in accordance with the Russian Civil Construction Standard CN - 25/74, which establishes a limit of C_A> 0.65. In Brazil, there are still no standards that set limits for the coefficient of resistance to water. These results are shown graphically in Figure 4.11, which shows the strong influence of the AEO percentage on the reduction in water resistance.

Composition 04 with C_A = 0.95, as in the uniaxial strength results (Table 4.9), showed higher mechanical property values, i.e. even saturated, the strength decreased by only 5% compared to CP in ambient conditions.

TABLE 4-9 - RESULTS OF THE WATER RESISTANCE TESTS ON THE 28TH DAY OF CURING.

No	Compositions, mass %			Water resistance on the 28th day		
	FG	AEO	RPC	R_{AMB}	R_{SAT}	C_A
1	60	25	15	7,6	6,0	0,79
2	70	15	15	5,6	4,6	0,82
3	65	15	20	7,1	6,1	0,86
4	**60**	**20**	**20**	**9,8**	**9,3**	**0,95**
5	55	30	15	7,7	6,9	0,90
6	50	40	10	7,8	7,1	0,91
7	45	45	10	6,3	5,3	0,84
8	40	45	15	5,2	4,0	0,77
9	25	55	20	5,5	4,6	0,84
10	15	65	20	4,8	4,2	0,87
11	10	70	20	5,4	4,3	0,79
12	5	75	20	5,0	3,8	0,76

Figure 4.12 shows that the lowest water resistance values occur at lower RPC and FG contents and, consequently, at higher AEO contents. The highest resistance value is related to the lime content, between 21 and 22 per cent.

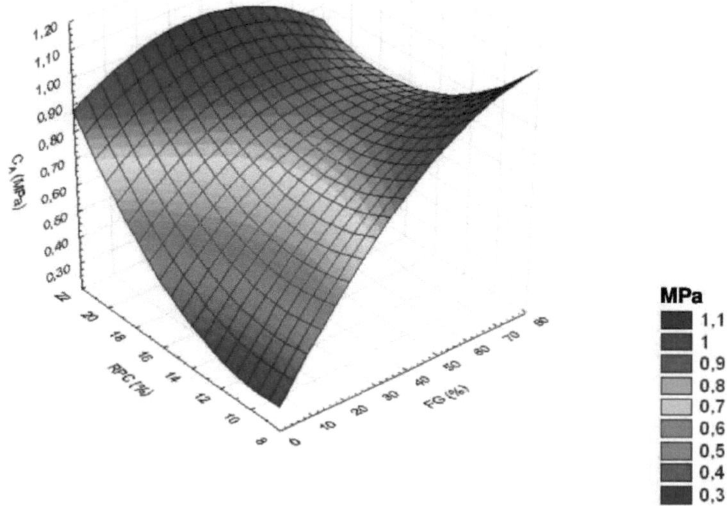

FIGURE 4-12 - VARIATION IN WATER RESISTANCE AT DIFFERENT FG AND RPC CONTENTS AT 28 DAYS OF CURING.

4.1.4.3 Water absorption

The results of the immersion water absorption test are shown in Table 4.10. According to NBR 8.953/1992, the maximum value allowed for average absorption in hollow concrete blocks for masonry with no structural function is 10%. Analysing the data in Table 4.10 shows that the highest water absorption coefficient values belong to the compositions with the highest binder content - 20% RPC. This means that during the 28 days of hydration and curing, all the lime was hydrated and involved in the process of forming the materials.Compositions 6 and 7 showed the lowest water absorption values, this result is probably due to the fact that these compositions have the lowest RPC contents.

TABLE 4-10- RESULTS OF THE CPS WATER ABSORPTION TEST AFTER 28 DAYS OF CURING.

No	Compositions, mass %			Dry weight (g)	Wet Weight (g)	CAA (%)
	FG	**AEO**	**RPC**			
1	60	25	15	11,58	12,56	8,47
2	70	15	15	11,71	12,69	8,36
3	65	15	20	11,65	12,65	8,56
4	**60**	**20**	**20**	**11,57**	**12,57**	**8,60**
5	55	30	15	11,46	12,38	8,00
6	50	40	10	11,44	12,28	7,69
7	45	45	10	11,41	12,25	7,38
8	40	45	15	11,33	12,29	8,47
9	**25**	**55**	**20**	**11,30**	**12,29**	**8,76**
10	15	65	20	11,24	13,72	8,56
11	10	70	20	11,29	12,24	8,45

12	5	75	20	11,18	12,12	8,44

Figure 4.13 shows the variation in the water absorption coefficient as a function of the RPC content, but there was no relationship between the FG content and the CAA. The relationship between the water coefficient and the RPC content can be explained by the granulometry of the raw material, which possibly absorbs more water.

The relationship between the water absorption coefficient and the RPC content can be explained by the granulometry of the raw material, given that 67 per cent of the RPC had a diameter of less than 0.18 mm.

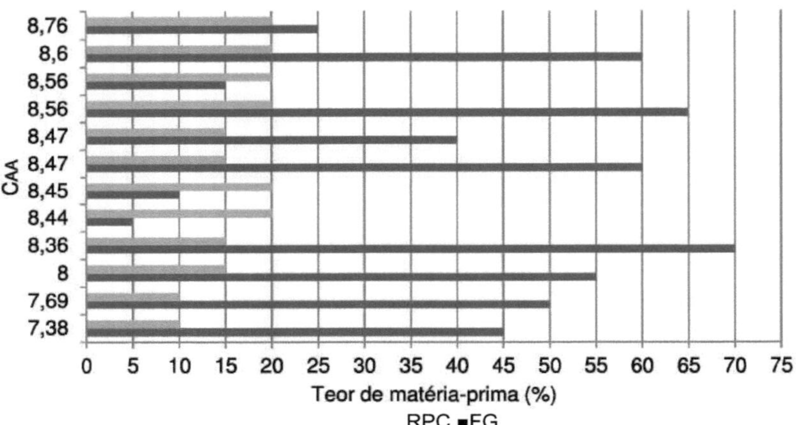

FIGURE 4-13 - VARIATION IN WATER ABSORPTION COEFFICIENT AND RPC AND FG CONTENT AFTER 28 DAYS OF CURING.

4.1.4.4 Expansion of materials during curing

The change in the coefficient of expansion of the CPs, shown in Table 4.11, with 10%, 15% and 20% lime production residue (RPC) during hydration and curing shows the increase in the coefficient of expansion of the materials at most curing times. The coefficient of expansion increases almost rectilinearly with 3 days of curing, from 0.54% to 1.01%, probably due to the increase in RPC content, with the decrease in phosphogypsum content, the coefficient showed values of 1.34% to 1.94%, after 1.5 years of curing.

As for the dimensional variations recorded up to 28 days of curing, it was possible to observe that compositions 6 and 7, which have a lower RPC content (10%), showed less expansion of the samples. Under the same conditions, compound 4 showed some tendency to expand between the 7th and 28th day of curing, and after this period it showed less tendency towards dimensional stability.

TABLE 4-11 - CHANGES IN THE COEFFICIENT OF EXPANSION OF THE SAMPLES DURING CURING.

Ns	Compositions, mass %			Expansion coefficient (%) in relation to curing time								
	FG	AEO	RPC	3 days	7 days	14 days	28 days	60 days	90 days	180 days	1 year	1.5 years
1	60	25	15	0,75	1,1	1,24	1,05	0,83	0,5	0,63	0,46	1,35
2	70	15	15	0,82	0,97	1,14	1,26	1,39	1,48	1,51	1,62	1,6
3	65	15	20	0,89	1,07	1,12	1,15	1,16	1,2	1,24	1,33	1,34
4	60	20	20	1	1,07	1,33	1,4	1,49	1,41	1,42	1,51	1,5
5	55	30	15	0,85	1,25	1,21	1,1	0,84	0,93	1,28	1,3	1,3
6	50	40	10	0,54	0,64	0,78	0,81	0,93	0,87	1,19	1,17	1,15
7	45	45	10	0,58	0,63	0,77	0,89	0,97	1,16	1,28	1,39	1,35
8	40	45	15	0,8	1,2	1,38	1,35	1,23	1,35	1,4	1,38	1,4
9	25	55	20	0,95	1,17	1,29	1,35	1,29	1,36	1,44	1,57	1,55
10	15	65	20	1,01	1,29	1,25	1,23	1,38	1,45	1,58	1,63	1,6
11	10	70	20	0,93	1,17	1,38	1,56	1,75	1,82	1,83	1,9	1,82
12	5	75	20	0, 96	1,11	1,24	1,32	1,39	1,57	1,67	1,88	1,94

Figure 4.14 shows the variation in the coefficient of expansion for different levels of phosphogypsum and lime production waste after 1.5 years of curing.

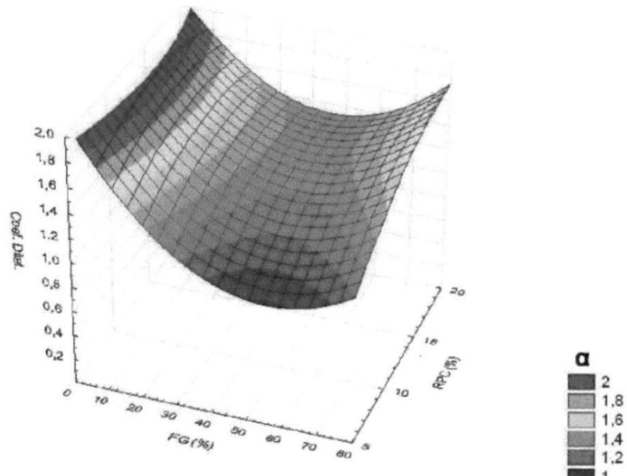

FIGURE 4-14 - VARIATION IN THE COEFFICIENT OF EXPANSION AS A FUNCTION OF THE COMPOSITION OF FG AND RPC, AFTER 1.5 YEARS OF CURING.

The graph in Figure 4.14 shows the inversely proportional relationship between the expansion coefficient and the GF content, where the lower the GF content, the higher the expansion coefficient.

4.2 Physical and Chemical Processes in the Formation of New Materials

The variation in mechanical properties mentioned above is due to the processes of new formations during the hydration and curing of the initial mixtures.

Composition 04, which showed the best mechanical properties, was used to study the physicochemical processes involved in the formation of structures. The changes in mineralogical composition, structure and chemical composition during the curing of the material were evaluated in three phases: Initial Dry Mix, 3 days and after 1.5 years of curing.

4.2.1 Changes in mineralogical composition during curing

The diffractograms in Figures 4.15 and 4.16 characterise the change in the mineralogical composition of composition 4 during hydration and curing of the initial dry mix and after 1.5 years, the last curing age studied according to Table 4.12.

TABLE 4-12 - COMPARISON OF THE MINERALOGICAL COMPOSITIONS OF THE INITIAL MIX SAMPLES AND AFTER 18 MONTHS OF CURING.

Initial dry mix (Figure 4.15)			18 months of curing (Figure 4.16)		
N°	Minerals	Chemical Formula	N°	Minerals	Chemical Formula
1	Albita	InAlSiaOa	1	Albita	IN THE SISTS
2	Plaster	$CaSO_42H_2O$	2	Plaster	$CaSO_42H_2O$
3	Dolomite	CaMg(CO3)2	3	Dolomite	CaMg(CO3)2
4	Magnesium Oxide Carbonate	Mg3O(CO3)2	4	Magnesium Oxide Carbonate	$Mg3O(CO3)_{(2)}$
5	Aluminium Calcium Magnesium Fluorsilicate	CaMg6Al2Sie02oF 4	5	Magnesium Fluoride Nitrate	MgsNFa
6	Quartz	SiO_2	6	Quartz	SiO_2
7	Muscovy	$K,Na(Al,Mg,Fe)_2$ $(Sĺ3.ıAlo.9)Oıo(OH)_2$	-	-	-
8	Microcline	KAISĺ3O8	-	-	-
9	Anhydride	$CaSO40,67H_2O$	-	-	-

Comparing the mineralogical compositions, it was possible to observe the disappearance of the anhydride. This behaviour was due to hydration, a process in which the chemical reaction between the anhydrous material and water occurs, regenerating the dihydrate, according to the chemical reaction described in equation 2.1 (HINCAPIÉ; CINCOTTO, 1995).

The crystallisation stage can be seen, when the solution becomes supersaturated. You can also see the dihydrate crystals (CaSO4.2H2O), where the particles precipitate in the form of needles and, after these stages, the mechanical phenomenon of hardening occurs, as the concentration of the crystals increases and the paste hardens.

Analysis and comparison of the diffractograms show that the peaks of lime, muscovite, microcline and calcium magnesium aluminium fluorsilicate have disappeared. Alumino-silicates are unstable in alkaline and humid environments, as is the case with the composition studied. It is clear that the crystalline structures have been partially or totally eroded in this environment and the products have been absorbed by the new formations.

The fluorine ions and part of the magnesium ions were used to synthesise magnesium fluoride nitrate, while the other portion of magnesium was possibly used to synthesise and permeate the crystalline structures of dolomite and magnesium oxide carbonate, as confirmed by the increase in the peaks shown.

The Albite crystals decreased, probably because they turned into gypsum, and the Quartz peaks remained stable after curing. The growth of amorphous substances during the curing process can be seen at the bottom of the diffractogram. The increase in amorphous substances of different chemical composition and density over time occurs through the dissolution of the solid components of the liquid phase in the pores of the material. Compacting the pores of amorphous substances can influence the increase in the material's resistance.

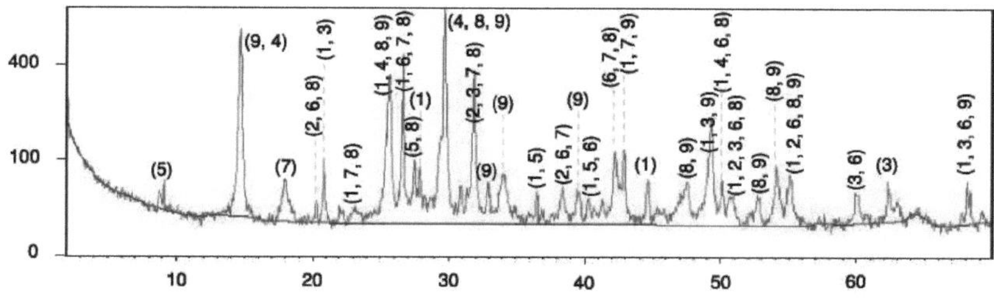

FIGURE 4-15 - DIFFRACTOGRAMS OF THE INITIAL DRY MIXTURE OF COMPOSITION 4.

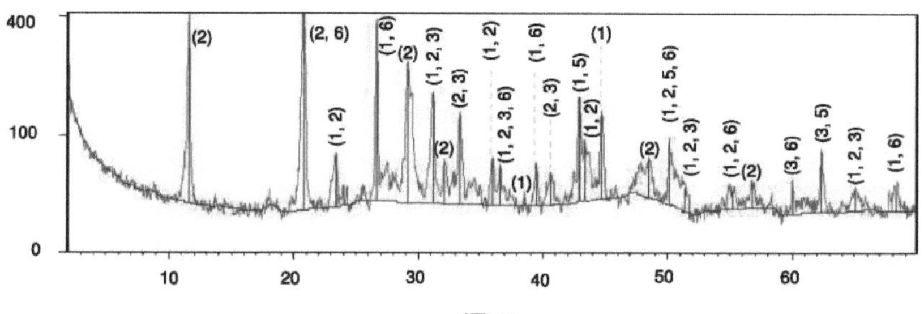

FIGURE 4-16 - DIFFRACTOGRAMS OF COMPOSITION 4 AFTER 1.5 YEARS OF CURING.

4.2.2 Changes in material structures during curing

4.2.2.1 Initial dry mix

The morphology of the composition 4 Initial mixture was checked using SEM and the EDS method was used to study the chemical composition of the material, as shown in Table

4.13.

In the micro-images, Figure 4.17, a general area was selected, and then 2 areas and 4 points were selected at the other magnifications for EDS.

The micro-images in Figure 4.17 show the mixing of the raw materials; as they were neither hydrated nor compacted at this stage, the particles showed no plastic deformation. In the micro-image in Figure 4.17 - D, the particle with a hexagonal structure, typical of phosphogypsum, stands out. The other particles do not have clear shapes to identify them.

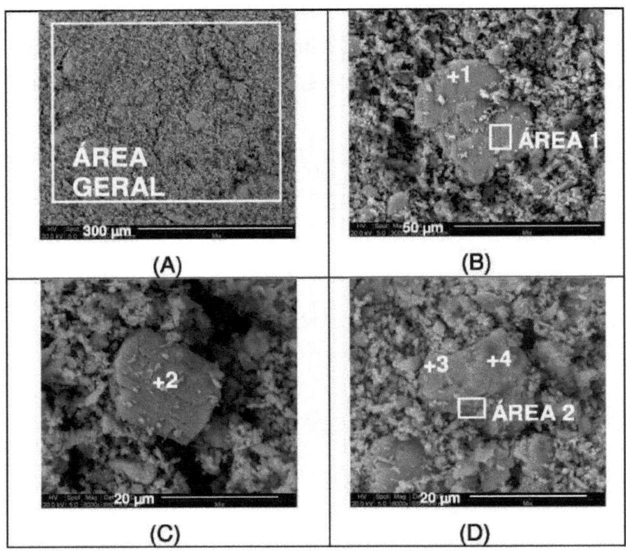

FIGURE 4-17- MICROIMAGE (MEV) OF THE INITIAL DRY MIXTURE AND EDS POINTS.

The chemical composition of the initial dry mix in the General area, shown in Figure 4.17 - A, is made up of various elements, including Calcium, Sulphur, Silicon and Carbon, with the highest content. In area 1 of Figure 4.17 - B and points 1, 3 and 4, Figure 4.17 - B and D, the chemical composition is similar to phosphogypsum. Area 2, Figure 4.17 - D and point 2 in Figure 4.17 - C show the contribution of gold mining sand particles on the surface of the phosphogypsum crystal.

Table 4.13 shows the results of the general chemical composition of the Initial Dry Mix, which is mostly composed of Calcium (45.91%), Sulphur (21.53%) and Carbon (14.36%), which gives the material better mechanical properties. In smaller concentrations, it contains Magnesium (6.02%), Aluminium (1.82%), Silicon (8.00%), Potassium (1.57%) and Iron (0.08%). This composition is characteristic of a mixture of phosphogypsum, gold mining sand and lime production waste.

TABLE 4-13 - RESULTS OF THE EDS ANALYSIS - CHEMICAL COMPOSITION OF THE AREAS AND POINTS OF THE INITIAL DRY MIX.

Spectrum	Chemical composition (EDS) of the Initial Dry Mix area, relative %									
	C	F	Mg	Al	Yes	S	K	Ca	Fe	Total
General	14,36	-	6,02	1,82	8,00	21,53	1,57	45,91	0,80	100,00
Area 1	13,68	-	0,39	-	-	36,54	-	49,38	-	100,00
Area 2	15,34	3,25	0,86	8,19	64,04	-	5,67	1,30	1,35	100,00
Point 1	12,54	-	-	-	-	36,96	-	50,51	-	100,00
Point 2	18,68	-	0,60	0,78	75,79	-	1,14	3,01	-	100,00
Point 3	14,14	-	0,40	-	-	35,75	-	49,71	-	100,00
Point 4	11,00	-	-	-	-	37,53	-	51,47	-	100,00

All the points and areas determined have compositions typical of each material: when the sulphur and calcium content is more evident, as in points 3 and 4, it is probably phosphogypsum particles; when the silicon content is more evident, as in the case of point 2, it is gold mining sand; and when the calcium content is more evident, point 1, it is lime production waste particles.

4.2.2.2 Material after 3 days of curing

The morphology of the specimens after 3 days of curing was studied using the Scanning Electron Microscopy (SEM) method and the results are shown in Figure 4.18. The chemical composition was studied using the EDS method, the results of which are shown in Table 4.14.

The micro-images in Figure 4.18 show the mixing of the raw materials after hydration, but as there is still only 3 days of curing time, the particles have little chemical interaction with each other.

As with the initial dry mix, the micro-image in Figure 4.18 - D shows the hexagonal particle characteristic of phosphogypsum.

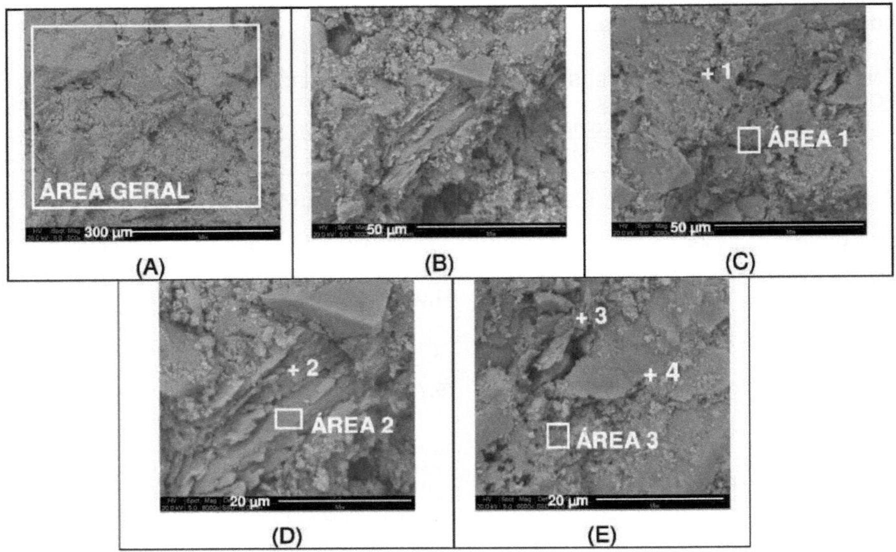

FIGURE 4-18 - MICROIMAGE (MEV) OF CP - 3 DAYS AND EDS POINTS.

The chemical composition found in the general area of the micro-image showed a mixture of all the components used as raw materials

TABLE 4-14- RESULTS OF THE EDS ANALYSIS - CHEMICAL COMPOSITION OF THE CP AREAS AND POINTS - 3 DAYS.

Spectrum	Chemical composition (EDS) of the CP area - 3 days, relative %										
	C	F	In	Mg	Al	Yes	S	K	Ca	Fe	Total
General	1,58	22,99	-	0,64	9,77	3,41	12,60	10,50	1,59	36,92	100,00
Area 1	18,94	-	6,96	2,15	13,81	53,27	-	0,76	5,03	-	100,00
Area 2	21,81	-	1,23	11,94	3,02	9,56	13,06	0,36	38,26	0,77	100,00
Area 3	17,31	-	-	28,68	0,66	0,87	8,25	-	43,31	0,92	100,00
Point 1	29,95	-	-	2,13	-	63,21	-	-	4,71	-	100,00
Point 2	8,67	-	-	-	-	-	34,10	-	57,23	-	100,00
Point 3	14,94	-	-	1,14	-	0,86	31,56	-	51,50	-	100,00
Point 4	14,94	-	-	1,14	-	0,86	31,56	-	51,50	-	100,00

According to Table 4.14, the majority of Area 1 and Point 1 contain Silicon, which indicates that they are Gold Extraction Sand particles. Points 2, 3 and 4 have considerable levels of sulphur and carbon, which characterises phosphogypsum particles. Areas 2 and 3, due to their high levels of Carbon and Calcium, are probably particles of Lime Production Waste.

4.2.2.3 Material after 1.5 years of healing

The specimens of the mixture of phosphogypsum, gold mining sand and lime production waste, after 1.5 years of curing, were subjected to scanning electron microscopy (SEM) to study their morphology, as shown in Figure 4.19.

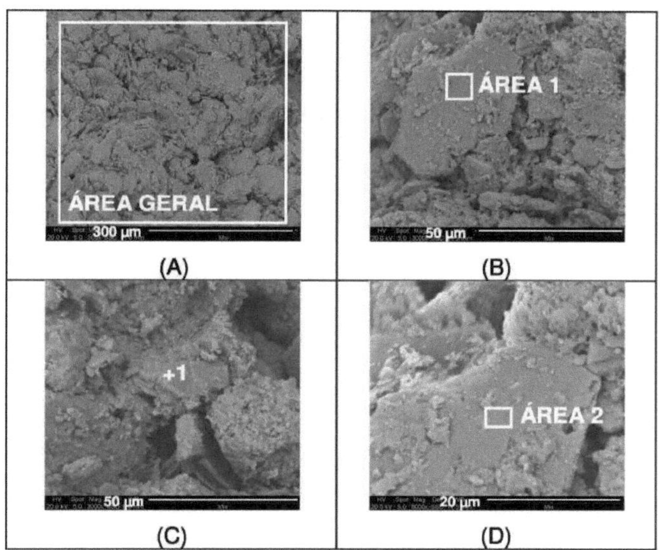

FIGURE 4-19 - MICROIMAGE (MEV) OF CP - 1.5 YEARS AND EDS POINTS.

In the micro-image in Figure 4.19 - A, the crystalline formations and the difference in sizes and configurations between the particles are visible, becoming more so with increased magnification. Different particle shapes can be seen: angular, laminar, needles, curved, among others.

The presence of pores of different sizes and configurations and the occurrence of plastic deformation increasing the size of the particles, probably due to slippage, were observed among the particles.

The chemical composition of the specimen after 1.5 years of curing was determined using the EDS method, shown in Table 4.15.

It can be seen that the composition of the general area at CP 1.5 years shows the highest levels of Calcium (35.59%), Carbon (31.15%) and Sulphur (12.33%). To a lesser extent, there was Silicon (8.71%), Magnesium (8.32%), Aluminium (2.35%) and Potassium (1.56%), with Area 1 showing similar characteristics to the general area. The presence of Potassium in Point 1 was possibly due to contamination of the sample. Area 2 showed Calcium (43.30%) and Carbon (29.66%) contents, which is typical of phosphogypsum.

TABLE 4-15 - RESULTS OF THE EDS ANALYSIS - CHEMICAL COMPOSITION OF THE CP AREAS AND POINTS - 1.5 YEARS.

Spectrum	Chemical composition (EDS) of the CP area - 1.5 years, relative %							
	C	Mg	Al	Yes	S	K	Ca	Total
General	31,15	8,32	2,35	8,71	12,33	1,56	35,59	100,00
Area 1	24,55	11,83	0,70	2,25	13,91	-	46,86	100,00
Area 2	29,66	18,12	0,68	0,63	7,50	-	43,30	100,00
Point 1	26,30	1,37	-	-	28,55	43,78	-	100,00

Compared to the EDS results for the initial dry mix, there was a considerable increase in Carbon, which can be explained by the fact that the mortar is prepared by mixing the FG with the RPCs, AEO and water, which produces a thick mass that loses water and solidifies when exposed to air, then gradually hardens as it absorbs carbon dioxide, forming calcium carbonate, which is used to form new structures and improve mechanical properties. Calcium remained in the same proportions due to the gypsum and dolomite minerals that remained in the material.

CHAPTER 5

CONCLUSION

Phosphogypsum, together with residual sand and lime production waste (badly burnt lime), can be used as the main raw material for new construction materials, reducing the risk of environmental impacts by properly disposing of this waste.

The materials developed had a strength of 8.3 MPa on the 3rd day of curing and reached 13.5 MPa after 90 days of curing. They also had a water resistance coefficient of 0.95 and water absorption of 8.76% after 28 days of curing. The results of the mechanical properties of the materials met the requirements for uniaxial compressive strength set out in NBR 7.170/83 from the third day onwards.

After 1.5 years of curing, it was possible to observe the occurrence of plastic deformation increasing the size of the particles, probably due to their slipping.

By studying the physical-chemical processes of the components of the initial mixtures and the formation of the new materials, the increase in mechanical property values during curing was due to the dissolution in an alkaline environment of the minerals from the gold mining sand (Microcline and Muscovite), the minerals present in the phosphogypsum (Calcium Magnesium Anhydride and Aluminium Fluorsilicate), with the synthesis of Magnesium Fluoride Nitrate and a large portion of the amorphous and crystalline carbonates.

The objectives of this research did not include economic efficiency calculations, but the use of three industrial wastes to replace natural raw materials is usually economically viable due to the low cost of waste compared to traditional raw materials.

The biggest beneficiary of the results obtained in this work is the environment, considering the real possibility of using industrial waste as a raw material, applying an appropriate final destination, preventing possible contamination of the environment and, above all, minimising the extraction of natural resources due to the reuse of this waste. The biggest challenge for production on an industrial scale is studying the market in order to attract entrepreneurs to invest in this new product.

SUGGESTIONS FOR FUTURE WORK

Following the conclusion of the research carried out in this paper, the following topics are suggested for future work:

i. Development of appropriate technology to scale up research results on an industrial scale;
ii. The evaluation of the mechanical properties of the final products (blocks, bricks, slabs, etc.) manufactured with the composition studied here.
iii. The study of the economic viability of manufacturing new materials from the 04 composition.

BIBLIOGRAPHICAL REFERENCES

AQUINO, P. E. **The production of phosphoric acid and the generation of phosphogypsum.** In: Technological Challenges for the Reuse of Phosphogypsum, 1, 2005. Belo Horizonte. Electronic proceedings... Belo Horizonte: UFMG, 2005. Available at:<http:/www.fosfogesso.eng.ufmg.br>. Accessed on: 21 January 2011.

ABREU, S. F. **Mineral resources of Brazil.** Rio de Janeiro, vol.1, *2nd* edition, 1965.

ANGELIN, R. R.; ANGELIN, R. C. M.; CARASEK, H.. **Influence of the granulometric distribution of sand on the behaviour of mortar coatings.** V Brazilian Symposium on Mortar Technology. São Paulo. 2003.

ARAÚJO, R.C.L.; RODRIGUES, E. H. V.; FREITAS, E. G. A. **Materiais de Construções - Coleção Construções Rurais.** Editora Universidade Rural. Rio de Janeiro. 2000.

ARMAN, A.; SEALS, R. K. **A preliminary assessment utilisation alternative for phosphogypsum.** Third International Symposium on Phosphogypsum. p. 562-575. Orlando, 1990.

BRAZILIAN ASSOCIATION OF TECHNICAL STANDARDS. **NBR 12.118** - Simple hollow concrete blocks for masonry - Test methods. 2011.

BRAZILIAN ASSOCIATION OF TECHNICAL STANDARDS. **NBR 12.129 -** Plaster for construction - Determination of mechanical properties. 1991.

BRAZILIAN ASSOCIATION OF TECHNICAL STANDARDS. **NBR 13.207 -** Plaster for civil construction - Requirements. 1994.

BRAZILIAN ASSOCIATION OF TECHNICAL STANDARDS. **NBR 13.529 -** Inorganic mortar wall and ceiling coverings. 1995.

BRAZILIAN ASSOCIATION OF TECHNICAL STANDARDS. **NBR 9.935** - Aggregates - Terminology. 2011.

BRAZILIAN ASSOCIATION OF TECHNICAL STANDARDS. **NBR 9.778** - Hardened mortar and concrete - Determination of water absorption, void index and specific masses. 2009.

BRAZILIAN ASSOCIATION OF TECHNICAL STANDARDS. **NBR 8953-** Concrete for structural purposes - Classification by specific mass, strength and consistency groups. 1992.

BRAZILIAN ASSOCIATION OF TECHNICAL STANDARDS. **NBR 8.492** - Solid soil-cement brick - Determination of compressive strength and water absorption - Test method. 1984.

BRAZILIAN ASSOCIATION OF TECHNICAL STANDARDS. **NBR 7.211** - Aggregates for concrete - Specification. Amendment 1:2009. 2005.

BRAZILIAN ASSOCIATION OF TECHNICAL STANDARDS. **NBR 7.213** - Lightweight aggregates for thermal insulating concrete. 1984.

BRAZILIAN ASSOCIATION OF TECHNICAL STANDARDS. **NBR 7.170** - Solid ceramic brick for masonry. 1983.

BRAZILIAN ASSOCIATION OF TECHNICAL STANDARDS. **NBR 6.453** - Quicklime for civil construction - Requirements. 2003.

BRAZILIAN ASSOCIATION OF TECHNICAL STANDARDS. **NBR 5.739** - Concrete - Compression test of cylindrical specimens. 1994.

BRAZILIAN ASSOCIATION OF TECHNICAL STANDARDS. **NBR 2.395** - Test sieve and sieving test - Vocabulary. 1995.

BECKER, P. **Phosphates and phosphoric acid: raw materials, technology and economics of the west process.** Fert Science Technology Service.V.6 p. 752. New York, 1989.

BHATTY, J. I.; GAJDA, J. **Alternative materials.** World Cement, v 35, n 12, p 41-48, Dec. 2004.

BORMA, S. SIMONE; SOARES, S. M. PAULO. **Gold Extraction - Principles, Technology and the Environment: Acid drainage and management of solid mining waste.** CETEM - Mineral Technology Centre. Rio de Janeiro, 2002.

BRODKOM, F. **Good environmental practices in the extractive industry: a reference guide.** Mines and Quarries Division of the Geological and Mining Institute, 2000. Available at http://eGeo.ineti.pt/geociencias/edicoes_online. Accessed on 20 January 2010.

CAHETÉ, S. FREDERICO. **Gold mining in the Amazon and its implications for the environment.** Federal University of Pará. Pará. 1995.

CALLISTER Jr, W. **Materials Science and Engineering: an Introduction.** 5§. Edition. Editora LTC. 2002.

CAMPOS, M. IBERÊ. **Sand for construction: how to buy and how to use it.** Available at: http://www.forumdaconstrucao.com.br /conteudo.php?a=31&Cod=44. Accessed on: 12/03/2011.

CANUT, M. MARIANA. **Physico-chemical characterisation of phosphogypsum waste.** Department of Materials and Construction. Federal University of Minas Gerais. P. 01-17. Belo Horizonte, 2005.

CARNEIRO, A. M. P.; CINCOTTO, M. A. Mortar **dosage using granulometric curves.** Technical Bulletin no. 237. Polytechnic School of the University of São Paulo. São Paulo. 1999.
CEKINSKI, E. **Fertiliser Production Technology.** Institute of Technology, 1990.

CHANG, W.F.; MANTELL, M.L ***Engineering properties and construction applications of phosphogypsum.*** Florida Institute of Phosphate Research. Coral Gables: Florida,. 201 p. ISBN 87024-28-4. 1990.

CINCOTTO, M. A., AGOPYAN, V. and FLORINDO, M. C. **Gypsum as a building material. Building Technology - Part I.** IPT-PINI, p.53-56. São Paulo, 1988.

CONAMA - NATIONAL ENVIRONMENTAL COUNCIL. **CONAMA Resolution No. 307 of July 2002.** Establishes guidelines, criteria and procedures for the management of construction waste. 2002.

CORRÊA, S.; Mymrine, V. A.. **Composite based on concrete waste and lime production waste.** Postgraduate Programme in Engineering (PIPE). Federal University of Paraná - UFPR. Curitiba. 2005.

CUNHA, P. JEFFERSON. **Development of a new material from the composition of varvito mining waste and lime production.** Postgraduate Programme in Engineering - PIPE. Federal University of Paraná. Paraná. 2007.

DRAGO, C; VERNEY.J. C.; PEREIRA, F. M. Effect of using sand from crushing in Portland cement concrete. Revista Escola de Minas. Ouro Preto. 2009.

DEõIRMENCI, N. **Utilisation of phosphogypsum as raw and calcined material in manufacturing of building products.** Construction and Building Materials. Volume 22. Balikesir University. Turkey. 2007.

DNPM - National Department of Mineral Production. **Brazilian Mineral Summary 2008.** Ministry of Mines and Energy. Brasilia, 2008.

GARAY, ALEXANDRE. **SINDICAL PR - Union of lime industries in the state of Paraná: History of lime production.** Available at: http://www.fiepr.org.br/sindicatos/sindicalpr/FreeComponent3302content19867.shtml , accessed on: 03/03/2011.

GARCIA, ANA. **Cement and lime manufacturing sector.** General environmental inspectorate and spatial planning. Lisbon. 2008.

GUIMARÃES, J. E. P. **Dimensions of the Lime Universe.** In: V Encontro aberta da Indústria da Cal - o uso da cal na Engenharia Civil. São Paulo. 1985.

HINCAPIÉ, A.M.; CINCOTTO, M.A. **Effect of setting retardants on the hydration mechanism and microstructure of building plaster.** Ambiente construído, v.l.Sào Paulo, p.7-17.1997

KURYATNYK, T; ANGULSKI, C.; AMBROISE, J; PERA, J. **Valorisation of phosphogypsum as hydraulic binder.** Journal of Hazardous Materials. Volume 16O.Institut National des Sciences Appliqué es de Lyon.France. 2008.

LEMOS, Ângela D. **C.Cleaner production as a generator of competitiveness: the case of the Cerro do Tigre farm.** Porto Alegre, 1998, 182 p.. Master's Thesis - Federal University of Rio Grande do Sul.

LUDWIG, U, SINGH, N. **Hydration of hemidrate of gypsum and its supersaturation.** Cementand concrete research, v. 18, p. 191-300.1978.

MARQUES, M. ANA. **Aerial lime mortar with added rice husk ash. Influence of curing conditions.** Technical University of Lisbon - Military Academy. Lisbon. 2010.

MAZZILLI, P. B.; PALMIRO, V.; SAUEIA, C.; NISTI, M. B. **Radiochemical characterisation of Brazilian phosphogypsum.** Journal environmental radioactivity. p. 113-122. 2000.

MAZZILLI, P. BÁRBARA. **Technological challenges for the reuse of phosphogypsum: The radioactivity of phosphogypsum.** Institute for Energy and Nuclear Research. Environmental radiometry laboratory. São Paulo. 2005.

MINEROPAR - MINERAIS DO PARANÁ S.A. **Mining masterplan for the Metropolitan Region of Curitiba.** Curitiba. 2004.

MYMRIN, V.A., DRAGOWSKY A., KACZYNSKI R, WORONKIEVICH S.D.. **The role of CaO in the ashes of thermal power stations utilisation as construction materials.** Actual problems of Engineering Geology, pp.73-84, Warsaw. Poland. 1975.

MYMRIN V.A., WORONKEVITCH **S.D.The effect of carbonate slime content of construction properties of thermal power stations ashes.** Proceedings of second International Congress of International Association of Engineering Geology, v. 1, pp. 315-319, São Paulo, Brazil.1974

NUERNBERG, N. J.; RECH, T. D.; BASSO, C. **Use of agricultural gypsum.** Technical Bulletin No. 112. Empresa de Pesquisas Agropecuária e Extensão Rural de Santa Catarina S.A. (EPAGRI). Santa Catarina, 2005.

PATTON, W.J. **Construction materials for engineering.** 2nd edition. E.P.U. Editora Pedagógica e Universitária. São Paulo. 1976.

PEREIRA, S. LUANA. **The lime industry in Brazil.** Mineral Technology Centre. Rio de Janeiro. 2009.

PETRUCCI, E. G. R. **Materiais de Construção.** 2-edition. Editora Globo. 1976.

PORTO, G. CLÁUDIO; PALERMO, NELY; PIRES, FERNANDO. **Gold Extraction - Principles, Technology and the Environment: Overview of gold exploration and production in Brazil.** CETEM - Mineral Technology Centre. Rio de Janeiro, 2002.

RIBEIRO, C. C.; PINTO, J. D. S.; STARLING, T. **Materiais de construção civil.** Editora UFMG. 2002.

RUTHERFORD, P. M. ; DUDAS, M. J. ; SAMEK, R. A. **Environmental impacts of phosphogypsum.** The Science total environmental, v. 149, p. 1-38,1996.

SABBATINI, F. H. **Pathology of coating mortars - physical aspects.** Proceedings: 3rd National Symposium on Construction Technology. Polytechnic School of the University of São Paulo. São Paulo. 1986.

SANTOS, A. J. G. **Avaliação do impacto radiológico ambiental do fosfogesso brasileiro e lixiviação de 226Ra e 210Pb**. Tese (Doutorado) - Instituto de Pesquisas Energéticas e Nucleares. São Paulo, 2002.

SCANDOLARA, J. P. **Properties of mortars obtained by partially replacing portland cement with particulate brick waste.** Dissertation (Master's Degree in Materials Science and Engineering). Santa Catarina State University. Joinville. 2010.

SEBBAHI, S.; CHAMEIKH, M.L.O.; SAHBAN, F.; ARIDE, J.; BENARAFA,L.; BELKBIR, L. **Thermal behaviour of Moroccan phosphogypsum.** *Elsevier: Thermochimica Acta,* n. 302, p.69-75. Morocco. 1997.

SENES, CONSULTANTS LIMITED. **An analysis of the major environmental and health concerns of phosphogypsum tailings in Canada and methods for their reduction.** Canada, 1987.

SHICHIERI, P. S.; PABLOS, J. M.; FERREIRA, O. P.; ROSSIGNOLO, J. A. and CARAM, R. **Construction Materials** I: **Mineral Binders, Aggregates, Mortars, Concrete and Dosage.** EESCUSP. São Paulo. 2008.

SILVA, N. G. **Lime and crushed limestone sand coating mortar.** Dissertation (PGCC Master's Degree). UFPR. 2006.

SINGH, M.; GARG, M.; REHSI, S. **Purifying phosphogypsum for cement manufacture.** Construction and Building Materials. Volume 7. Central Building Institute. India. 1993.

SINGH, M. **Treating waste phosphogypsum for cement and plaster manufacture.** Cement and Concrete Research. Volume 32. Central Building Research Institute. India. 2002.

SMADI, M.M. HADDAD, R.H. AKOUR, A.M. **Potential use of phosphogypsum in concrete.** *Elsevier: Cement and Concrete Research,* n. 29, p. 1419-1425, Jordan University of Science & Technology. Jordan. 1999.

STANDARDISED CONSTRUCTION METHODS. **CN - 25-74 -** Construction materials. Moscow. 1974.

STROEVEN, P.; VU, D.D.; BUI, D.D.; DONG, A.V. **Research on cementitious materials to promote sustainable developments in Vietnam.** In: International conference on concrete and development, 1, Tehran, 2001. Proceedings. [s.l.] : Building and Housing Research Centre, 2001.

THORMAN, C. H.; Dewitt, E.; Maron, M. A. C.; Ladeira, E. A. **Major Brazilian Gold Deposits.** Mineralium Deposita. 2001.

ULMANSS. **Ulmanss's encyclopedia of industrial chemistry.5** ed. Weinheim. Federal Republic of Germany: s 1 p., s. d.

VALVERDE, M. FERNANDO. **Aggregates for civil construction. SINDIMINERAL -** Pará **Mineral Industries Union.**Balanço Mineral Brasileiro. 2001.

VILLAVERDE, L. FREDDY. **Evaluation of external exposure in a residence built with phosphogypsum.** Institute for Energy and Nuclear Research. University of São Paulo. São Paulo, 2008.

YANG, J; LIU, W; ZHANG.L.; XIAO, B. **Preparation of load-bearing building materials from autoclaved phosphogypsum.Construction and** Building Materials.Volume 23.Huazhong University of Science and Technology. China. 2009.

ANNEX - EDS SPECTRA

PHOSPHOGYPSUM EDS

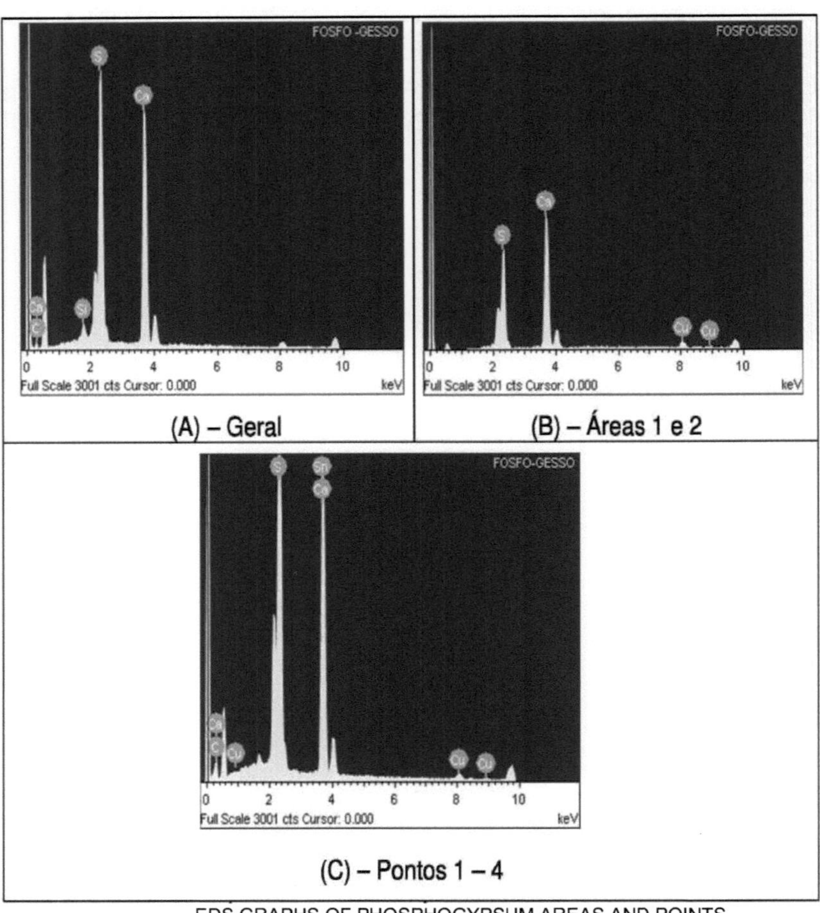

EDS GRAPHS OF PHOSPHOGYPSUM AREAS AND POINTS

EDS OF GOLD EXTRACTION SAND

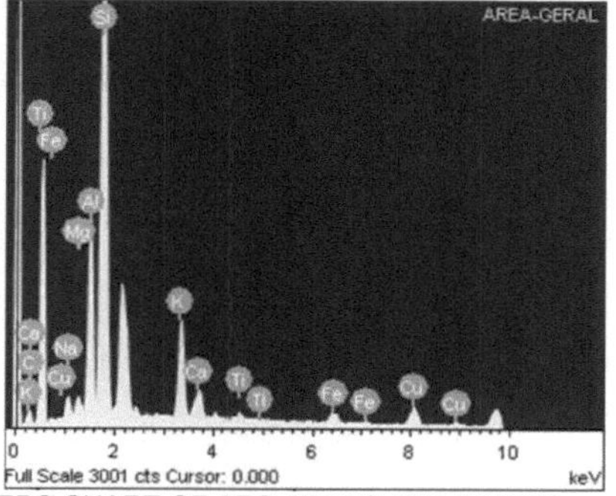

EDS CHART OF AEO AREAS AND POINTS

EDS OF THE INITIAL DRY MIX

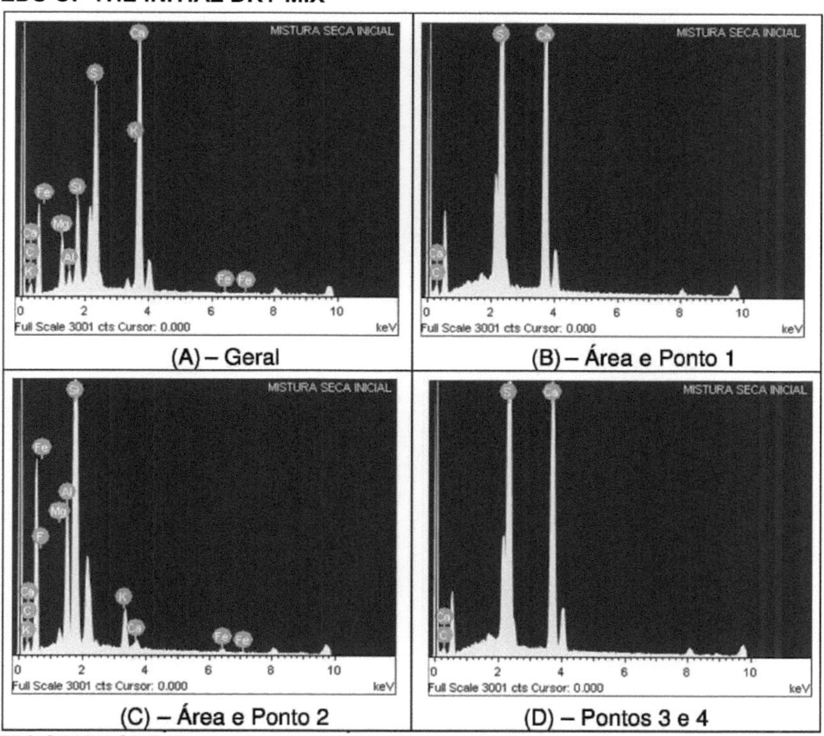

(A) – Geral

(B) – Área e Ponto 1

(C) – Área e Ponto 2

(D) – Pontos 3 e 4

EDS GRAPH OF THE AREAS AND POINTS OF THE INITIAL DRY MIX

EDS OF THE MATERIAL AFTER 3 DAYS OF CURING

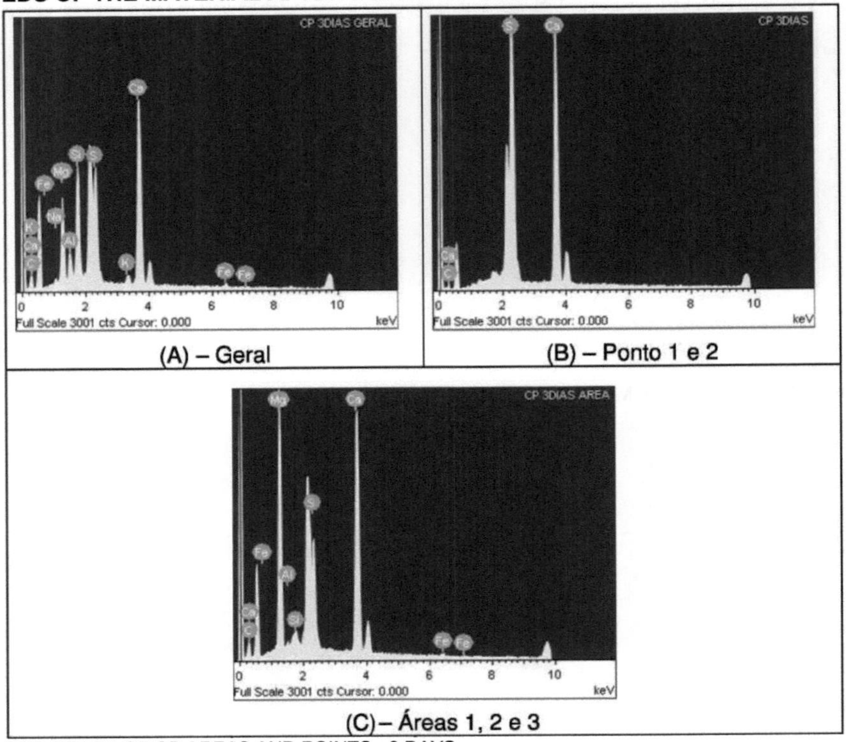

(A) – Geral

(B) – Ponto 1 e 2

(C)– Áreas 1, 2 e 3

EDS GRAPH OF CP AREAS AND POINTS - 3 DAYS

EDS OF THE MATERIAL AFTER 1.5 YEARS OF CURING

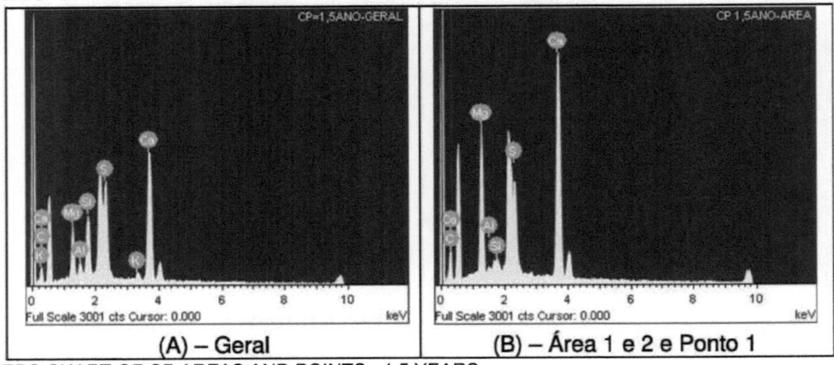

(A) – Geral

(B) – Área 1 e 2 e Ponto 1

EDS CHART OF CP AREAS AND POINTS - 1.5 YEARS

MIX
Papier aus verantwortungsvollen Quellen
Paper from responsible sources
FSC® C105338

Printed by Books on Demand GmbH, Norderstedt / Germany